T0200691

The Transformation of Mathematics in the Early Mediterranean World: From Problems to Equations

The transformation of mathematics from ancient Greece to the medieval Arab-speaking world is here approached by focusing on a single problem proposed by Archimedes and the many solutions offered. In this trajectory Reviel Netz follows the change in the task from solving a geometrical problem to its expression as an equation, still formulated geometrically, and then on to an algebraic problem, now handled by procedures that are more like rules of manipulation. From a practice of mathematics based on the localized solution (and grounded in the polemical practices of early Greek science) we see a transition to a practice of mathematics based on the systematic approach (and grounded in the deuteronomic practices of Late Antiquity and the Middle Ages). With three chapters ranging chronologically from Hellenistic mathematics, through Late Antiquity, to the medieval world, Reviel Netz offers a radically new interpretation of the historical journey of pre-modern mathematics.

REVIEL NETZ is Associate Professor in the Department of Classics at Stanford University. He has published widely in the field of Greek mathematics: *The Shaping of Deduction in Greek Mathematics: A Study in Cognitive History* (1999) won the Runciman Prize for 2000, and he is currently working on a complete English translation of and commentary on the works of Archimedes, the first volume of which was published in 2004. He has also written a volume of Hebrew poetry and a historical study of barbed wire.

CAMBRIDGE CLASSICAL STUDIES

THE TRANSFORMATION OF MATHEMATICS IN THE EARLY MEDITERRANEAN WORLD: FROM PROBLEMS TO EQUATIONS

REVIEL NETZ

CAMBRIDGE UNIVERSITY PRESS
Cambridge, New York, Melbourne, Madrid, Cape Town, Singapore,
São Paulo, Delhi, Dubai, Tokyo, Mexico City

Cambridge University Press
The Edinburgh Building, Cambridge CB2 8RU, UK

Published in the United States of America by Cambridge University Press, New York

www.cambridge.org
Information on this title: www.cambridge.org/9780521829960

First published 2004
Reprinted 2005, 2006

A catalogue record for this publication is available from the British Library

Library of Congress Cataloguing in Publication data
The Transformation of Mathematics in the Early Mediterranean World : From
Problems to Equations / Reviel Netz.
p. cm. – (Cambridge classical studies)
Includes bibliographical references.
ISBN 0 521 82996 8
1. Mathematics – Europe – History. I. Title. II. Series.
QA27.E85N48 2004
510′.94 – dc22 2003060601

ISBN 978-0-521-82996-0 Hardback
ISBN 978-0-521-04174-4 Paperback

To Maya and Darya

CONTENTS

ACKNOWLEDGMENTS

My words of gratitude are due, first of all, to the Classics Faculty at Cambridge, where I followed Sir Geoffrey Lloyd's lectures on Ancient Science. I remember his final lecture in the class – where "finality" itself was questioned. Just what makes us believe science "declined" at some point leading into Late Antiquity? Do we really understand what "commentary" meant? Do we not make a false divide between Greek and Arabic science?

Such questions rang in my ears – and in the many conversations Lloyd's students have had. Serafina Cuomo, in particular, helped me then – and since – to understand Late Ancient Science.

In this book I return to these questions and begin to offer my own reply. I am therefore especially grateful to the Faculty of Classics for allowing this book to be published under its auspices, in the Cambridge Classical Studies series.

My luck extends beyond the Faculty of Classics at Cambridge. In Tel Aviv, my first teacher in Greek mathematics was Sabetai Unguru, who has opened up to me the fundamental question of the divide between ancient and modern mathematics. Here at the Department of Classics, Stanford, I enjoy an exciting intellectual companionship and a generous setting for research. In particular, I work where Wilbur Richard Knorr once lived: and my sense of what an ancient problem was like owes everything to Knorr's research.

I am also grateful to Jafar Aghayani Chavoshi, Karine Chemla, David Fowler, Alain Herreman, Jens Hoyrup, Ian Mueller, Jacques Sesiano and Bernard Vitrac for many useful pieces of advice that have found their way into the book. Most of all I thank Fabio Acerbi who has read the manuscript in great detail, offering important insight throughout. In particular, Acerbi has generously shared with me his research on the *epi*-locution (the subject of section 2.5). I mention in my footnotes the references suggested by Acerbi, but

I should add that while my interpretation differs from Acerbi's, it is now formulated as a response to his own research (which I hope to see published soon). Needless to say, none of the persons mentioned here is responsible for any of my claims or views.

Some of the argument of Chapter 2 has been published as an article at *Archive for History of Exact Sciences* 54 (1999): 1–47, "Archimedes Transformed: the Case of a Result Stating a Maximum for a Cubic Equation," while some of the argument of Chapter 3 has been published as an article at *Farhang (Institute for Humanities and Cultural Studies, Tehran)* 14, nos. 39–40 (2002): 221–59, "Omar Khayyam and Archimedes: How does a Geometrical Problem become a Cubic Equation?" I am grateful, then, for permission to re-deploy some of the arguments of these articles, now in service of a wider goal: understanding the transformation of mathematics in the early Mediterranean.

INTRODUCTION

Does mathematics have a history? I believe it does, and in this book I offer an example. I follow a mathematical problem from its first statement, in Archimedes' *Second Book on the Sphere and Cylinder*, through many of the solutions that were offered to it in early Mediterranean mathematics. The route I have chosen starts with Archimedes himself and ends (largely speaking) with Omar Khayyam. I discuss the solutions offered by Hellenistic mathematicians working immediately after Archimedes, as well as the comments made by a late Ancient commentator; finally, I consider the solutions offered by Arab mathematicians prior to Khayyam and by Khayyam himself, with a brief glance forward to an Arabic response to Khayyam.

The entire route, I shall argue, constitutes history: the problem was not merely studied and re-studied, but transformed. From a geometrical problem, it became an equation.

For, in truth, not everyone agrees that mathematics has a history, while those who defend the historicity of mathematics have still to make the argument. I write the book to fill this gap: let us consider, then, the historiographical background.

My starting point is a celebrated debate in the historiography of mathematics. The following question was posed: are the historically determined features of a given piece of mathematics significant to it *as mathematics*? This debate was sparked by Unguru's article from 1975, "On the Need to Re-write the History of Greek Mathematics".[1] (At its background, as we shall mention below, was the fundamental study by Klein, from 1934–6, *Greek Mathematical Thought and the Origins of Algebra*.)

[1] For a survey of the debate, see Unguru (1979), Fried and Unguru (2001) and references there.

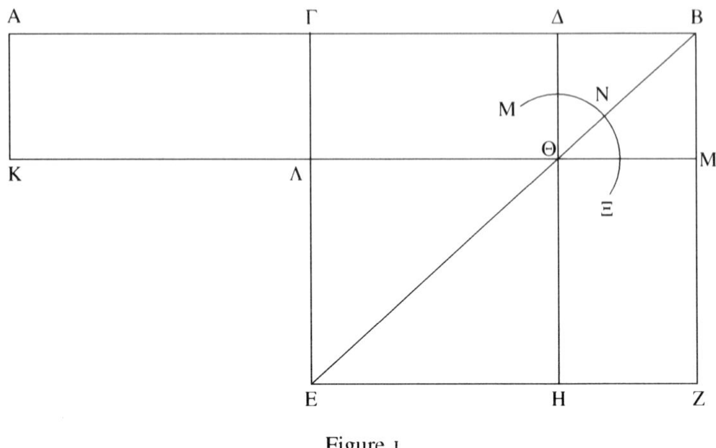

Figure 1

At the heart of Unguru's article was a simple claim for historicity. Theorems such as Euclid's *Elements* II.5, "If a straight line is cut into equal and unequal <segments>, the rectangle contained by the unequal segments of the whole, with the square on the <line> between the cuts, is equal to the square on the half" (see fig. 1) were read, at least since Zeuthen (1886), as equivalent to the modern equation $(a + b)(a - b) + b^2 = a^2$. That Euclid had not referred to any general quantities, but to concrete geometrical figures; that he did not operate through symbols, but through diagrams; and that he reasoned through manipulations of the rectangles in the diagram, cutting and pasting them until the equality was obtained – all this was considered, by authors such as Zeuthen, as irrelevant. As a pure mathematical structure, the equivalence between Euclid's formulation and modern algebra is straightforward. It is also indeed true that, for the modern reader, the best way to ascertain the validity of Euclid's theorem is by correlating it with the symbolic notation. And here arrives the seduction of a-historicism: mathematics is supposed to be compelling, it overpowers its readers by the incontrovertibility of its arguments. So, the a-historicist feels, unless one is overpowered by the argument, it is not really mathematical. The real form of the mathematical argument, then, is the form through which the reader feels its validity – that is, for a modern reader, the modern form. In its geometrical cloth, the Euclidean formulation is rendered inaccessible to the modern reader, so that it is no longer, for him or her, a piece of mathematics. Zeuthen considered himself within his rights, then, in removing the dust of the ages and uncovering the real form of Euclid's theorem, which was, according to Zeuthen, algebraic. The Greeks merely clothed their algebra geometrically, so that we may call this type of science "geometrical algebra." According to authors such as Zeuthen, all the historian of science needed to mark, as historian of science, was that, in the field of algebra, the Greeks had obtained such equations as $(a + b)(a - b) + b^2 = a^2$. That they then clothed these equations in geometrical form belongs not so much to the history of science itself, but to the history, so to speak, of scientific dresses: the sartorics of mathematics.

It was against this a-historical view that Unguru cried aloud in his article from 1975. At the time, he became the target of attack from some of the most distinguished historians of mathematics. A little over a quarter-century later, it is already difficult to doubt the basic correctness of Unguru. The exercise of geometrical algebra appears, in retrospect, as a refutation through absurdity of the a-historicist approach to mathematics. There are many reasons for this, but the most important is perhaps the following. By transforming the geometrical relations of *Elements* II into an algebraic equation, they are rendered trivial: so that, instead of allowing us to see better the significance of ancient argument, we, instead, lose sight of its importance for the ancient audience. The moral seems to be that, if, indeed, the way to understand a mathematical text *as mathematical* is by perceiving its validity; and if indeed the perception of validity depends on historically conditioned tools (e.g., diagrams, for the Greeks, symbols, for the moderns) – then the way to understand ancient mathematics is not by transforming it into our mathematical language but, on the contrary, by becoming, ourselves, proficient in the mathematical language of the ancients. The skill of parsing arguments through diagrams is as essential to a historical understanding of Euclid as the skill of parsing Greek hexameter is essential to a historical understanding of Homer.

All of which, however, still does not get us into *history*. While most historians of mathematics would now agree on the need to understand mathematical texts through the language of their times, this amounts, so to speak, to a dialectology of mathematics, not to its history. Greek mathematics, granted, is different from modern mathematics: but what is the historical transformation that led from the first to the second?

The very success of Unguru, in challenging the old model of the a-historical "Geometrical Algebra," makes the problem more acute. For an a-historical scholar such as Zeuthen, there was no significant process beyond accumulation, so that the historian merely needed to record the dates and names involved with this, essentially static process. History, for Zeuthen, did not change; it merely boarded the escalator of progress. But what if the very nature of mathematics had changed with time? In this case, there is a complicated process characterizing the history of mathematics, and the first task of the historian would be to uncover its dynamics. But no convincing account has yet been offered of this process, so that Unguru's claim remains, at best, as a tantalizing observation, and, at worst, as a dogmatic statement of a gap between the ancient and modern "minds." For here is the paradox: unless some specific historical account is offered of the difference between ancient and modern mathematics, Unguru's claim can seem to be saying that *the ancients are just different from us and that is it*. In this way, we have come full circle to a-historicism, the single monolith "Mathematics" now broken into the two smaller a-historical monoliths, "Ancient" and "Modern."

Why did Unguru not offer such a historical account? This perhaps may be answered by looking for his historiographical ancestry. Indeed, the very assumptions that led Unguru to criticize geometrical algebra, also led him away from studying the dynamics of the transformation from the ancient to the modern. Unguru's premise was that of a great divide, separating ancient from modern thinking. The assumption of a great divide, in itself, is not conducive to the study of the dynamics leading from one side of the divide to the other. But more than this: Unguru's assumption of a great divide was, in turn, adopted from Klein's study *Greek Mathematical Thought and the Origins of Algebra* – which still remains the best statement of the difference between ancient and modern mathematics. It was Klein's study, specifically, that led scholars away from studying the dynamics of the transformation from the ancient to the modern. As it were: a-historical readers required no dynamics, while historical readers were satisfied with its absence, relying on the methodology offered by Klein. We should therefore turn briefly to discuss this methodology. But I should immediately emphasize that my aim now is not to argue against Klein. On the contrary, I see my book as a continuation and corroboration of Klein's thesis. It is, however, by seeing the shortcoming of Klein's approach, that a way beyond him could be suggested. I shall therefore concentrate in what follows on the shortcoming of this work which, undoubtedly, remains a study of genius.

Klein's approach went deeper than the forms of mathematics. For Klein, it was not merely that the ancients used diagrams while the moderns use symbols. To him, the very objects of mathematics were different. The ancients referred to objects, directly, so that their arithmetic (the case study Klein took) was a study of such objects as "2," "3," "4," etc. The moderns, however, refer to symbols that only then, indirectly, refer to objects. Thus modern arithmetic is not about "2," "3," and "4," but about "k," "n," and "p," with all that follows for the forms of mathematics. Ancient mathematics (and science in general) was, according to Klein, based on a first-order ontology; modern mathematics (and science in general) is based on a second-order ontology.

To repeat, my aim in this book is not to argue against Klein's main thesis, but rather to find a historical explanation for an observation that Klein offered mainly on a philosophical basis. However, it should be said that Klein's study was conceived in the terms of an abstract history of ideas that left little room for persuasive historical explanations. It is typical of Klein's methodology that he takes, as his starting point, not the mathematical texts themselves, but Plato's statements about mathematics, and that, inside mathematical texts, he is especially interested in methodological discussions and in definitions. When studying the history of arithmetic, Klein focuses on "the concept of the number." Klein's assumption is that, in different epochs, different concepts are developed. From the different fundamental distinction in concepts, the entire difference in the nature of the science follows.

I am not sure how valid this very approach to intellectual history is. I doubt, myself, whether any generalizations can be offered at the level of "the Greek concept of . . ." More probably, different

Greek thinkers had different views on such issues, as distinct from each other as they are from some modern views. Nor do I think that periods in the history of science are characterized by some fundamental concepts from which the rest follows. Sciences are not coherent logical systems, developed through an inert deduction from first principles: they are living structures, proceeding towards first principles, away from them, or, most often, in ignorance of them, revamping *ambulando* their assumptions. At any rate, regardless of what we think of Klein's method in general, it is clear that it made it very difficult for him to approach the dynamics of historical change. The neat divide, and its grounding in sharp conceptual dichotomies, simply left no room for a historical account of the transformation leading from the ancient to the modern. The issue was primarily a matter of logic, not of history. Klein merely sketched a possible account of this divide – and it is instructive to see the impasse that Klein had faced in this brief sketch (I quote from the English translation, Klein [1968] 120–1. All italics in the original):

> Now that which especially characterizes the 'new' science and influences its development is *the conception which it has of its own activity* . . . Whereas the 'naturalness' of Greek science is determined precisely by the fact that it arises out of 'natural' foundations [i.e. reference to the real world] . . . the 'naturalness' of modern science is an expression of its *polemical attitude towards school science*. In Greek science, concepts are formed in continual dependence on 'natural', prescientific experience . . . The 'new' science, on the other hand, generally obtains its concepts through a process of polemic against the traditional concepts of school science . . . No longer is the thing intended by the concept an object of *immediate* insight . . . In evolving its own concepts in the course of combating school science, the new science ceases to interpret the concepts of Greek *epistēmē* preserved in the scholastic tradition from the point of view of their 'natural' foundations; rather, it interprets them with reference to the function which each of these concepts has within the whole of science.

There is much in this paragraph that I find insightful, and I shall to a large extent adopt, in the following study, the basic distinction Klein offered between first-order concepts and second-order concepts. But notice how difficult it would be to sustain Klein's thesis, historically. Klein suggests: (a) that the main original feature of modern science is that it was polemical – as if Greek science was not! (b) that Greek science was throughout tied to pre-scientific, natural objects – whereas it was often based on flights of theoretical fancy, removed from any connection with the natural world; (c) finally, that somehow, by virtue of such differences, the Greeks would deal with "2," "3," and "4," while the moderns would deal with "k," "n," and "m" – how and why this follows, Klein cannot say.

The truth is that, aside from the shortcomings of the history of ideas as such – leading to Klein's emphasis on concepts, and to his ignoring practices – he was also a captive of certain received ideas about the basic shape of Mediterranean intellectual history, ideas that were natural in the early twentieth century but are strange to us today. "The Greeks," to him, were all of a piece (as were, of course, the "moderns"). History was told in terms of putting the first against the latter. What went in between was then twice misrepresented. First, it was reduced to the Latin Middle Ages (the "schools" Klein refers to), so that the most important medieval development of Greek science, in the Arab world, was ignored. Second – in part, as a consequence of the first – the Middle Ages were seen as a mere repository of ideas created in Antiquity, no more than rigidifying the past so that the modern world could rebel against the past's rigidity. Now try to offer an account of the path leading from point A to B, when you oversimplify the nature of points A and B, and then ignore, or misrepresent, what went in between them! It would be a piece of common sense that, if we want to understand the transformation separating antiquity from modernity, we should be especially interested in what went in between: Late Antiquity and the Middle Ages. To ignore them is simply to accept uncritically the false claim of modernity to have been born directly from the Classics. And it was a mere construct of European linguistic capacities, and prejudices, that had made enlightened scholars such as Klein ignore, effectively, Arabic civilization. In reality, no balanced picture of Mediterranean history can be offered, as it were, purely on the Indo-European.

The thesis of this book is that Classical Greek mathematics went through a trajectory of transformation through Late Antiquity and the Middle Ages, so that, in certain works produced in the Arab speaking world, one can already find the algebra whose origins Klein sought in modern Europe. The changes are not abrupt, but continuous. They are driven not by abstruse ontological considerations, but by changes in mathematical practice. To anticipate, my claim, in a nutshell, is that Late Antiquity and the Middle Ages were characterized by a culture of books-referring-to-other-books (what I call a *deuteronomic* culture). This emphasized ordering and arranging previously given science: that is, it emphasized the systematic features of science. Early Greek mathematics, on the other hand, was more interested in the unique properties of isolated problems. The emphasis on the systematic led to an emphasis on the relations between concepts, giving rise to the features we associate with "algebra." So that, finally, I do not move all that far from Klein's original suggestion: it was by virtue of becoming second order (though in a way very different from that suggested by Klein!) that Classical mathematics came to be transformed.

As mentioned at the outset, the following is a study of a single *case* of development, illustrating the transformation of early Mediterranean mathematics. Since I believe the process was driven not by conceptual issues, but by mathematical practice, I concentrate not, so to speak, on mathematics in the laboratory – definitions and philosophical discussions – but on mathematics in the field – that is, actual mathematical propositions. The best way to do this, I believe, is by following the historical development of a single mathematical proposition.[2] I take in this book a single ancient mathematical problem and study its transformation from the third century BC to the eleventh century AD – from geometrical problem to algebraic equation.

The book is informed by two concerns. First, I argue for the "geometrical" or "algebraic" nature of the problem at its various stages, refining, in the process, the sense of the terms. Second, I offer a historical account: why did the problem possess, at its different stages, the nature it possessed? The first concern makes a contribution to the debate on the historicity of mathematics, following Unguru, and my main aim there is to support and refine Unguru's

[2] In doing this, I also follow the methodology of Goldstein (1995).

position. The second concern aims to go beyond the historiographical debate and to give an explanation for the transformation of early Mediterranean mathematics.

In Chapter 1, I describe the nature of the problem within Classical Greek mathematics. Chapter 2 discusses the degree to which the problem was transformed in Late Antiquity, while Chapter 3 discusses its transformation in Arab science. The hero of Chapter 1 is Archimedes himself. In Chapter 2 the hero is Archimedes' commentator, Eutocius (though much mention is still made of Archimedes himself, so that the contrast between Archimedes and Eutocius can be understood). The hero of Chapter 3, finally, is Omar Khayyam, whose algebra is seen as the culmination of the trajectory followed here. Originally a problem, it now became an equation, and from geometry, algebra was created – leading, ultimately, to such authors as Zeuthen who would understand, retrospectively, Greek mathematics itself as characterized by a "geometrical algebra."

There are advantages and drawbacks to taking a single example. Most obviously, I open myself to the charge that my case study is not typical. My main thesis, that Late Antiquity and the Middle Ages were characterized by deuteronomic culture, with definite consequences for the practice of mathematics, was argued, in general terms, in an article of mine ("Deuteronomic Texts: Late Antiquity and the History of Mathematics," 1998). That article went through many examples showing the role of systematic arrangement in late Ancient and medieval mathematics. In this book I attempt a study in depth of a single case, and I shall not repeat here the examples mentioned in that article. But I should say something on this issue, even if somewhat dogmatically – if only so as to prevent the reader from making hasty judgments. For the reader might be surprised now: was not early Greek mathematics itself characterized by an interest in systematic arrangements? Two examples come to mind: that Ancient Greek mathematicians had produced many solutions to the same problem, leading to catalogues of such solutions; and that Ancient Greek mathematics had produced Euclid's *Elements*. As a comment to this I shall mention the following. First, the catalogues of ancient solutions are in fact the work of Late Ancient authors bringing together many early, isolated solutions.[3] Second, one may easily exaggerate the systematic nature of Euclid's *Elements* (I believe it is typical of early Greek mathematics that each of the books of the *Elements* has a character very distinct to itself: more on such deliberate distinctness in Chapter 1 below); but even so, I believe the work as we know it today may be more systematic than it originally was, due to a Late Ancient and Medieval transformation including, e.g., the addition of proposition numbering, titles such as "definitions" etc. Third, and most important, the centrality of Euclid's *Elements* in Greek mathematics is certainly a product of Late Antiquity and the Middle Ages – that had fastened on the *Elements* just because it was the most systematic of ancient Greek mathematical works. In early Greek mathematics itself, Euclid had a minor role, while center stage was held by the authors of striking, isolated solutions to striking, isolated problems – the greatest of them being, of course, Archimedes.[4]

This book is dedicated to what may be the most striking problem studied by Archimedes – so striking, difficult, and rich in possibilities, that it could serve, on its own, as an engine for historical change. Time and again, it had attracted mathematicians; time and again, it had challenged the established forms of mathematics. Quite simply, this is a very beautiful problem. Let us then move to observe its original formulation in the works of Archimedes.

[3] The most important such catalogue is Eutocius' survey of the solutions to the problem of finding two mean proportionals, in Heiberg (1915) 56–106.

[4] In this characterization of early Greek mathematics I follow Knorr (1986).

1

THE PROBLEM IN THE WORLD OF ARCHIMEDES

In this chapter I discuss the Archimedean problem in its first, "Classical" stage. In section 1.1, I show how it was first obtained by Archimedes and then, in 1.2, I offer a translation of the synthetic part of Archimedes' solution. Following that, section 1.3 makes some preliminary observations on the geometrical nature of the problem as studied by Archimedes. Sections 1.4 and 1.5 follow the parallel treatments of the same problem by two later Hellenistic mathematicians, Dionysodorus and Diocles. Putting together the various treatments, I try to offer in section 1.6 an account of the nature of Ancient geometrical problems. Why were the ancient discussions geometrical rather than algebraic – why were these *problems*, and not *equations*?

1.1 The problem obtained

In his *Second Book on the Sphere and Cylinder*, Archimedes offers a series of problems concerning spheres. The goal is to produce spheres, or segments of spheres, defined by given geometrical equalities or ratios. In Proposition 4 the problem is to cut a sphere so that its segments stand to each other in a given ratio. For instance, we know that to divide a sphere into two equal parts, the solution is to divide it along the center, or, in other words, at the center of the diameter. But what if want to have, say, one segment twice the other? Cutting it at two-thirds the diameter is clearly not the answer, and the question is seen to be non-trivial, for two separate reasons: it involves solid figures, and it involves curvilinear figures – both difficult to handle by simple manipulations of lines.

However, a direction forwards suggests itself. The two segments of the sphere share a common base – the plane at which they are divided – and certain solid and curvilinear figures are relatively easy to handle once their base is made equal: in particular, cones.

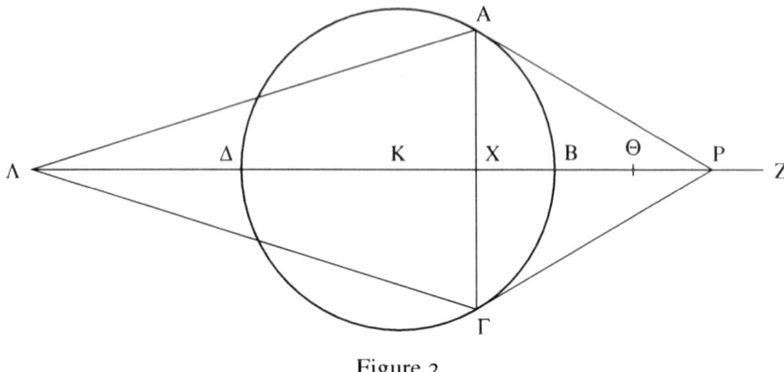

Figure 2

The ratio of cones of equal base is the same as the ratio of their height – in other words it is a simple linear ratio. Therefore we shall try to convert the segments of sphere into cones. This is obtainable following the measurement of segments of spheres, proved by Archimedes in Propositions 42 and 44 of his *First Book on the Sphere and Cylinder* (these serve as basis of the central theorem of the second book, II.2, where every segment of sphere is made equal to a cone). Hence the figure of this Proposition 4 (fig. 2): ΑΒΓ, ΑΔΓ are the two segments of sphere; ΑΡΓ, ΑΛΓ are the cones equal to them, respectively. The question "where to cut the sphere" is the question of the ratio between the diameter (ΒΔ) and one of the cut lines (e.g. ΔΧ). In the simplest case of equality, this ratio is 2:1, but in all other cases it still eludes us; but, with the cones, we have a way forwards.

Now, to get the cones, a relatively complex ratio defines the lines ΧΡ, ΧΛ in terms of the position of the point Χ. For instance, the length ΡΧ is defined by (transforming into a modern notation)

$$(\mathrm{K}\Delta + \Delta\mathrm{X}):\Delta\mathrm{X}::\mathbf{PX}:\mathrm{XB}.$$

Clearly, all lines except for ΡΧ are given by the point Χ itself, so that, in general, the cones are well defined and with them the ratio of the two segments of sphere. Thus a single manipulation by ratios, albeit a complex one, transforms a ratio defined by solid, curvilinear figures, into a ratio defined by lines alone.

Archimedes introduces now two auxiliary lines (that ultimately simplify the ratios). The line ΒΖ is defined in a simple way,

KB = BZ. As for the line ZΘ, it is defined in a more complex way: PΛ:ΛX::BZ:ZΘ. Notice however that while this ratio is somewhat complex, it is still "manageable," since the ratio PΛ:ΛX is essentially the ratio we would be given by the terms of the problem: it is the ratio of the sum of the cones (i.e. the sum of the segments of sphere, i.e. simply the sphere) to the smaller cone, i.e. the smaller segment, so if the problem is to cut the sphere in the ratio 2:1, the ratio PΛ:ΛX is 3:1. BZ, again, is simply the radius, so the point Θ is fully defined by the terms of the problem.

What happens now to the cutting-point itself, X? Our goal now is to manipulate our ratios so that we define the point X with the various lines we have defined by the terms of the problem. Archimedes reaches such a ratio:

(sq. on BΔ):(sq. on ΔX)::XZ:ZΘ.

In other words, the terms of the problem define a line ΔZ, and our task is to find a cutting-point on it, X. This cutting-point has a complex defining property.

The cutting-point cuts the line into two smaller lines, ΔX, XZ. Now, we have The Defining Square – the one on BΔ; and The Defining Line – ZΘ; both are fully determined by the terms of the problem. The Defining Property is this: The Defining Square has to the square on one of the smaller lines (ΔX) the same ratio which the other smaller line (XZ) has to The Defining Line.

The problem becomes truly irritating in its details if we continue to think about the specific characteristics of The Defining Area and The Defining Line, in terms of the problem. For instance, The Defining Area happens to be the square on two-thirds the given line ΔZ; while the definition of ZΘ is truly complex. It is much easier, then, simply to leave those details aside and to look at the problem afresh, without the specific characteristics: we can always reinsert them later when we wish to. So the problem can be re-stated as follows:

Let us assume we are given a line and an area – any line, any area. Let us re-name them, now, as the line AB and the area Δ. Now the problem is, given another line, which we call AΓ, to find a point on AB – say O – that defines two segments of AB, namely

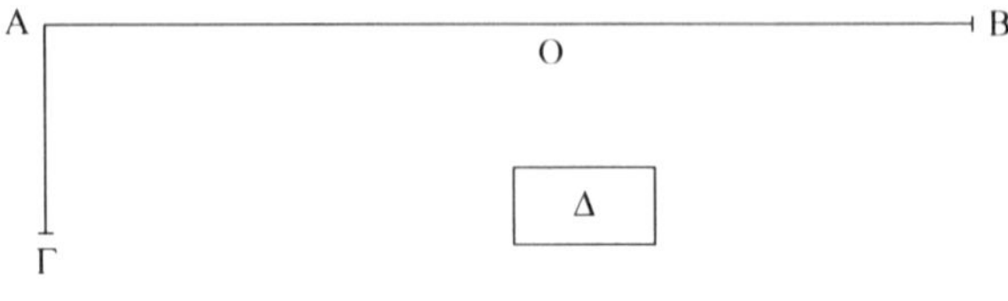

Figure 3

AO, OB. Those two segments should now satisfy:

$$AO:A\Gamma::(\text{area } \Delta):(\text{square on } OB).$$

This is the problem, whose evolution we study in this book.

Is the problem as stated now soluble? This is not yet evident. The point O serves, as it were, two masters: once, it defines AO, thus serving the ratio AO:AΓ; once again it defines OB, serving the ratio (area Δ):(square on OB). Can one be the servant of two masters? Yes, if the service is identical: the two ratios must be the same. It is as such – as a complex proportion – that Archimedes understands and solves the problem.

To take stock of the ground covered so far: the Archimedean problem arises directly from a well-defined geometrical task, of an immediate, "tangible" interest – to cut the sphere according to a given ratio. The problem is then transformed, and then solved, always following the principle of transforming geometrical ratios, until simple ratios between lines are obtained. At the moment where the ratios, while linear, become too complex to handle, Archimedes moves into a higher plane of generality, ignoring some specific properties of the problem at hand: but the purpose of this transition into generality is merely to arrive at ratios that are more simply defined. To be precise, Archimedes does not explicitly move into a higher plane of generality, since he can make use of the ambiguity of reference to the diagram. Here is Archimedes' own handling of the transition to the general problem:

> Therefore it is required to cut a given line, ΔZ, at the <point> X and to produce: as XZ to a given <line, namely> ZΘ, so the given <square, namely> the <square> on BΔ to the <square> on ΔX.

(At this point, Archimedes had reached the problem in its particular terms, the various lines referring to specific objects on the sphere. However, he goes on writing as if he were *already* referring to general lines):

> This, said in this way – without qualification – is soluble only given certain conditions, but with the added qualification of the specific characteristics of the problem at hand (that is, both that ΔB is twice BZ and that ZΘ is greater than ZB – as is seen in the analysis), it is always soluble; and the problem will be as follows:
>
> Given two lines BΔ, BZ (and BΔ being twice BZ), and <given> a point on BZ, <namely> Θ; to cut ΔB at X, and to produce: as the <square> on BΔ to the <square> on ΔX, XZ to ZΘ.

Archimedes reaches the general problem not like a schoolmaster, who tells us explicitly that a certain particular problem might also be conceived as a more general one. He reaches it, rather, like a conjurer. Having got us used all the while to thinking about a particular problem, suddenly he begins to talk about it as if it were *already* general – and suddenly, when we look back at the stage, we find that the protagonists ΔB, BZ, ZΘ, and ZB really are general – for there is nothing stopping us from looking at them in this way. We are not led into the general problem; we are surprised into it. This, I would say, is typical of Archimedes' style, trying to elicit from the audience the effect of awed surprise.[1]

This in itself is telling for Archimedes' approach: one thing in which he is clearly uninterested is the systematic explication of the relationship between general and particular presentations. On the contrary, Archimedes exploits the potential for ambiguity between the two, to obtain a specific rhetorical effect.

It still remains the case that Archimedes, as a matter of logic, does move into a higher plane of generality. As we shall see later on in the book, this move does lead us closer to what we may consider an "equation."

Here we reach, for the first time, a certain duality typical of Archimedes' approach to the problem. On the one hand, the problem is clearly embedded within a geometrical world, studying particular geometrical configurations. Generality is not an issue. On

[1] I have discussed this in detail in Netz (forthcoming), giving many examples from Archimedes' books *On Sphere and the Cylinder*, and have also described the same practice in Netz (2001), for other works as well. The most obvious example of this feature of Archimedes' rhetoric is indeed the overall structure of the *First Book on the Sphere and Cylinder*, best seen in *SC* 1.23.

the other hand, the complexity of the problem is such that it is required – just as a simplification – to effect a certain idealization of the problem, in this case casting it in a more general form. The genesis of the problem in Archimedes, as well as his cavalier way of introducing the general form of the problem, both suggest that, to him, this was really a problem about segments of spheres; yet, alongside this strict geometrical conception, there is also the possibility of a more general reading of the problem. What is especially interesting is that we need not ascribe to Archimedes any motivation to seek generalization for its own sake. Indeed, we need not even ascribe to him any awareness that, in its more generalized form, the problem has an interest transcending the case of the sphere. Rather, the very nature of the problem – briefly, its complexity – demands a simplification that, we now see, holds in it the germs of the abstract or indeed the algebraic. The trajectory, from problems to equations, is to a certain extent implied within the problem itself.

But we are pushing ahead. Let us first move on to read Archimedes' solution to the problem.

1.2 The problem solved by Archimedes[2]

And it will be constructed like this: let the given line be AB, and some other given <line> AΓ, and the given area Δ, and let it be required to cut the <line> AB, so that it is: as one segment to the given <line> AB, so the given <area> Δ to the <square> on the remaining segment.

(a) Let the <line> AE be taken, a third part of the <line> AB; (1) therefore the <area> Δ, on the <line> AΓ[3] is either greater than the <square> on BE, on the <line> EA, or equal, or smaller.

[2] The following is a translation of Heiberg (1915) 136.14–140.20. The argument that this text is indeed by Archimedes is not straightforward, but for the moment I shall assume the text read here is indeed by him. We shall return to discuss this in the next chapter, when considering the transformations of Archimedes in the world of Eutocius (through whom we have Archimedes' text). The translation is mine, and is adopted from Netz (forthcoming), where the conventions of translation are explained. Note, however, that (for reasons which will become apparent in Chapter 3 below, when I come to compare Archimedes' treatment with Khayyam's) I do not abbreviate the Greek expression "the <line> AB" into "AB," as I generally do in translations of Greek mathematics.

[3] The expression "area, on line" means "the parallelepiped with the area as base, and the line as height." We shall return to discuss this expression in the next chapter.

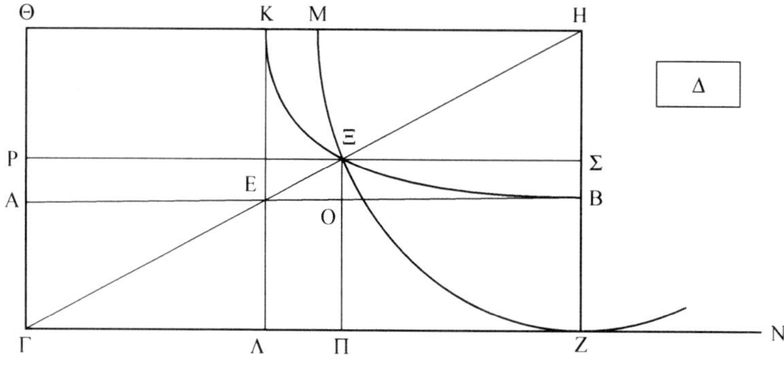

Figure 4

(2) Now then, if it is greater, the problem may not be constructed, as has been proved in the analysis;[4] (3) and if it is equal, the point E produces the problem. (4) For, the solids being equal, (5) the bases are reciprocal to the heights,[5] (6) and it is: as the <line> EA to the <line> AΓ, so the <area> Δ to the <square> on BE.

(7) And if the <area> Δ, on the <line> AΓ is smaller than the <square> on BE, on the <line> EA, it shall be constructed like this:

(a) Let the <line> AΓ be set at right <angles> to the <line> AB, (b) and let the <line> ΓZ be drawn through Γ parallel to the <line> AB, (c) and let the <line> BZ be drawn through B parallel to the <line> AΓ, (d) and let it meet the <line> ΓE (<itself> being produced) at H, (e) and let the parallelogram ZΘ be filled in, (f) and let the <line> KEΛ be drawn through E parallel to the <line> ZH. (8) Now, since the <area> Δ, on the <line> AΓ is smaller than the <square> on BE, on the <line> EA, (9) it is: as the <line> EA to the <line> AΓ, so the <area> Δ to some <area> smaller than the <square> on BE,[6] (10) that is, <smaller> than the <square> on HK.[7] (g) So let it be: as the <line> EA to the <line> AΓ, so the <area> Δ to the <square> on HM, (h) and let the <rectangle contained> by ΓZN be equal to the <area> Δ.[8] (11) Now since it is: as the <line> EA to the <line> AΓ, so the <area> Δ, that is the <rectangle contained> by ΓZN (12) to the <square> on HM, (13) but as the <line> EA to the <line> AΓ, so the <line> ΓZ to the <line> ZH,[9] (14) and as the <line> ΓZ to the <line> ZH, so the <square> on ΓZ to the <rectangle contained> by ΓZH,[10] (15) therefore also as the <square> on ΓZ to the <rectangle contained>

[4] The reference is to a later part of the same argument, showing the limits of solubility of the problem. We shall return to discuss this argument in the next chapter.

[5] *Elements* XI.34.

[6] The closest foundation in Euclid is *Elements* VI.16, proving that if $a*b = c*d$, then $a:d::c:b$ (for a, b, c, and d being lines).

[7] Steps b, e, f, *Elements* I.34.

[8] Steps g and h define the points M, N respectively, by defining areas which depend upon those points.

[9] Steps b, e, f, *Elements* I.29, 32, VI.4.

[10] *Elements* VI.1.

by ΓZH, so the <rectangle contained> by ΓZN to the <square> on HM;[11] (16) and alternately, as the <square> on ΓZ to the <rectangle contained> by ΓZN, so the <rectangle contained> by ΓZH to the <square> on HM.[12] (17) But as the <square> on ΓZ to the <rectangle contained> by ΓZN, the <line> ΓZ to the <line> ZN,[13] (18) and as the <line> ΓZ to the <line> ZN, (taking ZH as a common height) so is the <rectangle contained> by ΓZH to the <rectangle contained> by NZH;[14] (19) therefore also, as the <rectangle contained> by ΓZH to the <rectangle contained> by NZH, so the <rectangle contained> by ΓZH to the <square> on HM; (20) therefore the <square> on HM is equal to the <rectangle contained> by HZN.[15] (21) Therefore if we draw, through Z, a parabola around the axis ZH, so that the lines drawn down <to the axis> are, in square, the <rectangle applied> along the <line> ZN – it shall pass through M.[16] (i) Let it be drawn, and let it be as the <parabola> MΞZ. (22) And since the <area> ΘΛ is equal to the <area> AZ,[17] (23) that is the <rectangle contained> by ΘKΛ to the <rectangle contained> by ABZ,[18] (24) if we draw, through B, a hyperbola around the asymptotes ΘΓ, ΓZ, it shall pass through K[19] (through the converse of the 8th theorem of <the second book of> Apollonius' *Conic Elements*).[20] (j) Let it be drawn, and let it be as the <hyperbola> BK, cutting the parabola at Ξ, (k) and let a perpendicular be drawn from Ξ on AB, <namely> ΞOΠ, (l) and let the <line> PΞΣ be drawn through Ξ parallel to the <line> AB. (25) Now, since BΞK is a hyperbola (26) and ΘΓ, ΓZ are asymptotes,[21] (27) and the <lines> PΞΠ[22] are drawn parallel to the <lines> ABZ, (28) the <rectangle contained> by PΞΠ is equal to the <rectangle contained> by ABZ;[23] (29) so that the <area> PO, too, <is equal> to the <area> OZ. (30) Therefore if a line is joined from Γ to Σ, it will pass through O.[24] (m) Let it pass, and let it be as the <line> ΓOΣ. (31) Now, since it is: as the <line> OA to the <line> AΓ, so the <line> OB to the <line> BΣ,[25] (32) that is the <line> ΓZ to the <line> ZΣ,[26] (33) and as the <line> ΓZ to the <line> ZΣ (taking ZN as a common height) the <rectangle contained> by ΓZN to the <rectangle contained> by ΣZN,[27] (34) therefore as the <line> OA to the <line> AΓ, too, so the <rectangle contained> by ΓZN to the <rectangle contained> by ΣZN. (35) And the <rectangle contained> by ΓZN is equal to the area

[11] *Elements* v.11. [12] *Elements* v.16. [13] *Elements* vi.1.

[14] *Elements* vi.1. [15] *Elements* v.9. [16] The converse of *Conics* i.11.

[17] Based on *Elements* i.43.

[18] As a result of Step a (the angle at A right), all the parallelograms are in fact rectangles.

[19] Converse of *Conics* ii.12.

[20] This note was not put in by Archimedes, but by the later commentator Eutocius; interestingly – and typically – Eutocius' reference assumes a text of the *Conics* different from ours. For Eutocius' practices, particularly in regard to the *Conics*, see Decorps-Foulquier (2000).

[21] Steps 25–6: based on Step j.

[22] An interesting way of saying "the <lines> PΞ, ΞΠ." [23] *Conics* ii.12.

[24] Step 30 is better put as: "The diagonal of the parallelogram ΠΣZΓ passes through O," which can then be proved as a converse of *Elements* i.43.

[25] *Elements* i.29, 32, vi.4. [26] *Elements* vi.2. [27] *Elements* vi.1.

Δ,[28] (36) while the <rectangle contained> by ΣZN is equal to the <square> on ΣΞ, (37) that is to the <square> on BO,[29] (38) through the parabola.[30] (39) Therefore as the <line> OA to the <line> AΓ, so the area Δ to the <square> on BO. (40) Therefore the point O has been taken, producing the problem.

1.3 The geometrical nature of Archimedes' problem

The text quoted in the previous section will serve as the basis for comparison, when discussing the gradual transformation of Archimedes' problem into an equation. Right now, however, we are interested in a narrower question: was this problem, as conceived by Archimedes, at all geometrical? Why do we not consider it an "equation?"

It can be seen that, in his solution, Archimedes sometimes uses the remarkable expression, "the {area}, on the {line}," meaning something like a multiplication of an area by a line. This already seems to suggest a quasi-algebraic treatment of geometric objects (as if they were mere quantitative units, abstracted away from their spatial reality). This in itself is another case of the duality of Archimedes' solution – proceeding geometrically, but containing within it traces of the suggestion of an equation. We shall return to discuss this particular duality in Chapter 2 below.

Leaving this question aside for the moment, we may now return to Archimedes' text and reformulate it, using throughout the strange expression "the {area} on the {line}." In this way, we shall simplify the problem further. The following then is no longer Archimedes' own formulation of the problem, but it still does represent his mathematical tools. This simplification would be important when comparing Archimedes with later developments.

Recall the ratio obtained by Archimedes – the starting-point for the problem:

(sq. on BΔ):(sq. on ΔX)::XZ:ZΘ

Now, there being four lines in proportion, A:B::C:D, we deduce an equality between two rectangles:

(rectangle contained by A,D) equals (rectangle contained by B,C).

[28] Step h. The original Greek is literally: "To the <rectangle contained> by ΓZN is equal the area Δ" (with the same syntactic structure, inverted by my translation, in the next step).

[29] Steps a, e, k, l, *Elements* I.34.

[30] A reference to *Conics* I.11 – the "symptom" of the parabola.

While the extension of this result to parallelepipeds has a less compelling intuitive character, we just saw Archimedes taking it for granted in some moves of his solution. If applied on his formulation of the problem, then, it could have produced the following three-dimensional equality:

(parallelepiped cont. by sq. Δ X, line XZ) equals
(parallelepiped cont. by sq. BΔ, line ZΘ).

Now we can see that the bottom side is known – both square and line are given. Thus we are asked simply to cut a line so that the square on one segment, together with the other segment, produce a given parallelepiped. The seemingly intractable ratios of spheres, their segments and their cones, have been reduced to a truly elegant task.

Let us now translate the problem even further, now into modern terms, so as to have some vantage point from which, finally, to compare Archimedes with a more algebraic approach. So, the problem is that of cutting a line (call it a) so that the square on one of its segments (call this x^2) "multiplied" by the other segment ($a - x$) equals a certain given solid magnitude (call it b):

The square on a segment of a given line, "multiplied" by the remaining segment, equals a given magnitude, or

A cube, together with a given magnitude, equals a square "multiplied" by another given magnitude, or

$$x^2(a - x) = b, \quad \text{or} \quad x^2a - x^3 = b, \quad \text{or} \quad x^3 + b = x^2a.$$

This final re-formulation of the problem, as we shall see in Chapter 3 below, is clearly an equation of algebra. It is also highly reminiscent of a proposition in Khayyam's *Algebra*. In short, then, Archimedes' problem is equivalent to an algebraic equation – and, I shall try to argue below, it would later on transform into such an equation.

With Archimedes, however, it is not an equation, but a problem. This statement brings us right to the heart of the debate concerning the historicity of mathematics. So, to get us going, we need first to try and understand the mathematical meaning of the problem for Archimedes himself.

I shall now, first, sketch the line of reasoning that could have led Archimedes to his solution. I shall then contrast this sketch with a standard a-historicist account of this solution (offered by Heath in his *History of Greek Mathematics*). With this contrast in mind, we may begin to analyze the difference between problems and equations – and why this difference is important for the history of mathematics.

In fact, Archimedes' reasoning can be reconstructed quite straightforwardly, especially since we have a further clue to that. We had read above the synthetic part of a solution that Archimedes had presented inside an analysis-and-synthesis pair. In such a presentation, the problem is first assumed solved, in the "analysis," and certain conclusions are drawn from this assumption; those conclusions then suggest the required preliminary construction with the aid of which it is then possible, in the "synthesis," to solve the problem (this time, without *assuming* that it is already solved). It is well known that, in general, finding a useful analysis is just as difficult as finding a useful synthesis. Thus, merely reading the analysis does not provide us, necessarily, with an insight to the mathematician's process of discovery. Indeed, in this case, Archimedes' analysis itself involves special constructions, whose discovery would have been just as complicated as that of the solution to the problem itself. For one thing, already in the analysis, Archimedes constructs the parabola and the hyperbola, which are simply not given in the terms of the problem. In general, I had argued in Netz (2000) that the goal of the published versions of Greek mathematical analyses was largely expository, rather than heuristic. Analyses could serve to set out, in demonstrative form, the relations leading to the required construction. In short, then, we cannot simply use the analysis as a key to the heuristics guiding Archimedes in his search for the solution. We can use it, however, as an indication of the relations that, for Archimedes, underlay the solution. With this purpose in mind, I now quote from the analysis (Heiberg [1915] 132.23):

> Given a line, AB, and another, AΓ, and an area, Δ: let it first be put forth:[31] to take a point on AB, such as E, so that it is: as AE to AΓ, so the area Δ to the <square> on EB.

[31] I.e. "let the geometrical task be."

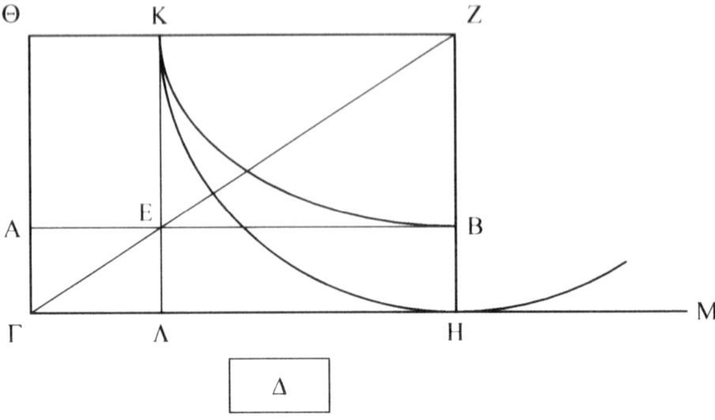

Figure 5

(a) Let it come to be, (b) and let AΓ be set at right <angles> to AB, (c) and, having joined ΓE, (d) let it be drawn through to Z, (e) and let ΓH be drawn through Γ parallel to AB, (f) and let ZBH be drawn through B parallel to AΓ, meeting each of the <lines> ΓE, ΓH, (g) and let the parallelogram HΘ be filled in, (h) and let KEΛ be drawn through E parallel to either ΓΘ or HZ, (i) and let the <rectangle contained> by ΓHM be equal to the <area> Δ.

(1) Now since it is: as EA to AΓ, so the <area> Δ to the <square> on EB,[32] (2) but as EA to AΓ, so ΓH to HZ,[33] (3) and as ΓH to HZ, so the <square> on ΓH to the <rectangle contained> by ΓHZ,[34] (4) therefore as the <square> on ΓH to the <rectangle contained> by ΓHZ, so the <area> Δ to the <square> on EB, (5) that is to the <square> on KZ;[35] (6) alternately also: as the <square> on ΓH to the <area> Δ, that is to the <rectangle contained> by ΓHM, (7) so the <rectangle contained> by ΓHZ to the <square> on ZK.[36] (8) But as the <square> on ΓH to the <rectangle contained> by ΓHM, so ΓH to HM;[37] (9) therefore also: as ΓH to HM, so the <rectangle contained> by ΓHZ to the <square> on ZK. (10) But as ΓH to HM, so (HZ taken as a common height) the <rectangle contained> by ΓHZ to the <rectangle contained> by MHZ;[38] (11) therefore as the <rectangle contained> by ΓHZ to the <rectangle contained> by MHZ, so the <rectangle contained> by ΓHZ to the <square> on ZK; (12) therefore the <rectangle contained> by MHZ is equal to the <square> on ZK.[39] (13) Therefore if a parabola is drawn through H around the axis ZH, so that the lines drawn down <to the axis> are in square the <rectangle applied> along HM,[40] it shall pass through K . . .[41]

[32] The assumption of the analysis. [33] *Elements* VI.2, 4, and I.34.
[34] *Elements* VI.1. [35] *Elements* I.34. [36] *Elements* V.16.
[37] *Elements* VI.1. [38] *Elements* VI.1. [39] *Elements* V.9.
[40] For any point Z on the axis, the square on the line drawn from the parabola to the point Z, i.e. the square on KZ, is equal to the rectangle contained by ZH (i.e. the line to the vertex of the parabola) and by the constant line HM (the *latus rectum*) – i.e. to the rectangle ZHM.
[41] Converse of *Conics* I.11.

Archimedes' line of argument is made to appear much more complicated than it really is, because he needs to express in detail the operations on proportions that yield his fundamental result. But the main thrust of the argument is very simple, and can be paraphrased as follows. We are given a complicated proportion: a given line is to be cut, so that one segment is to another given line, as a given area is to a square on the remaining segment.

Let us first move to give the first ratio (between the first segment and the given line) some concrete form. A ratio between lines can be expressed in terms of similar triangles within which they are embedded. For the sake of simplicity, it helps to make those triangles right-angled. We may then position the given line AΓ at right angles to the line AB, at the point A, and we reach the following pleasant relation: no matter where the cut E may fall, it will then define a ΓE across which, extended, the ratio between the lines EA, AΓ will always be conserved within a rectangle such as ZΘΓH. This is the thrust of the construction at Steps a–h.

At this point we may conceive of the line ΓZ, if we wish, as a sliding ruler, whose fixed point is at Γ, and whose point E slides along the line AB. (This conception has the advantage that it corresponds to several constructions in Hellenistic mathematics,[42] so that we are still using terms familiar to Archimedes himself). Now, we have already gained something: we have brought the other segment, EB, into play. The ratio of EA to AΓ is, e.g., the same as the ratio of EB to BZ (or, alternatively, of KZ to BZ – it helps that all the lines are either parallel or orthogonal, so that segments can be equated up and down, or sideways, automatically). Unfortunately, this is not quite what we seek: the problem specifies a ratio involving not the segment EB itself, but the square on it. We need some tool to equate the square on EB – or on KZ – with a given line segment.

Luckily, we have just the tool for that: parabolas directly transform ratios involving squares on lines, to ratios involving lines. If we assume a parabola passing through K and H, then, being a parabola, it will have the following property: the square on KZ is

[42] See, e.g., in Eutocius' commentary to Archimedes' *SC* II.2, the catalogue of solutions to the problem of finding two mean proportionals, the solutions by "Plato," Eratosthenes and Nicomedes.

equal to a rectangle, one of whose sides is HZ, while another side is an inert constant (say, we may call it HM). In other words, HZ acts, so to speak, as the linear representative of the square on KZ. A ratio involving the square on KZ is equivalent to a ratio involving the line HZ.

Now, at this point ΓZ is still a sliding ruler, so that the point K slides with it. There is thus not one parabola, but infinitely many. Which of them solves the problem? We need to take stock of the position, and consider which proportion we are actually interested in. What we would like the parabola to do, is to conserve:

AE:ΓA::(given area):(square on KZ), or

AE:ΓA::(given area):(rect. ZH, HM).

And it is at this point that the thought suggests itself: is it not useful, that the line ZH, too, participates in the ratio projected by the similar triangles? Instead of AE:ΓA, we can have ΓH:ZH! That is, we require

ΓH:ZH::(given area):(rect. ZH, HM).

But this is obviously fulfilled if the given area is equal to the rect. ΓH, HM (then, all we need to do to get from the first ratio to the second, is to add the common height of the two rectangles, HM). So this, in fact, is what we want: that the given area be equal to the rect. ΓH, HM.

But wait – we can arrange that! The point M is, as yet, unfixed; so we might as well fix it wherever we like. Let us then arrange that the point M falls where rect. ΓH, HM equals the given area, and we have defined a parabola solving, effectively, our problem.

The section I had quoted above from Archimedes' analysis corresponds, essentially, to the discursive account I had offered so far, and this is the heart of the solution: defining a point M that, in turn, defines the parabola that, in turn, equates a ratio between areas with a ratio between lines.

There is merely one remaining irritating feature: we still do not know where the point K is to be found! That is, the parabola already ensures that, once we have got everything surrounded by a rectangle, the correct proportions would follow. But how to get the rectangle? The point is this: with a given parabola such as a HK (given, now, because we have fixed the point M), the sliding ruler ΓZ would, generally, fail to encase the lines we need within a single rectangle. The sliding ruler ΓZ finds a point Z, which in turn defines, on the parabola, a point K, which in turn finds, on the line AB, a point E – but here the construction might collapse: the point E might fail to fall on the original sliding ruler ΓZ – might fail to fall on the diagonal line connecting the point Γ and Z.

How to ensure, then, that the point E falls on this diagonal? This is not as difficult as it seems. The fundamental property of a diagonal in a parallelogram, as discussed in Euclid's *Elements* (I.43), is that it keeps constant the equality between the small parallelograms erected on its two sides – wherever you take your point E, the areas ΘE, EH (or ΘΛ, AH) are equal to each other. So to ask that the point K would be such that, underneath it, the point E would fall on the diagonal ΓZ, is the same as asking that the point K would be such that the area ΘΛ would be equal to AH. With everything here being defined by right angles, the situation is even simpler: we wish a point K, so that the rectangle on ΘK, ΘΓ is equal to the rectangle ΓH, HB. We are seeking, then, some instrument for the preservation of the equalities between rectangles. No sooner said than done. We have just such a tool – the hyperbola, one of whose well-known features (e.g. in Apollonius' *Conics* in its present form, II.12) is that it may keep rectangles on asymptotes equal: a hyperbola passing at B, with its asymptotes ΘΓ, ΓH, would thereby preserve the equality (rect. BH, HΓ) = (rect. KΘ, ΘΓ). So the point K is clearly defined: it is at the intersection of the parabola obtained above with the hyperbola passing at B whose asymptotes are ΘΓ, ΓH.

This, then, seems to be the line of thought leading to Archimedes' solution, accounting for the basic pattern of the solution: a parabola and a hyperbola, embedded within a system of parallel or orthogonal lines.

Let us now contrast this with Heath (1921) II 43–5:

Cubic equation arising out of II.4.
. . . the generalized equation is of the form $x^2(a - x) = bc^2$ . . .
Archimedes's own solution of the cubic:

[this] is solved by means of the intersection of a parabola and a rectangular hyperbola, the equations of which may be written thus

$$x^2 = (c^2/a)y, \quad (a - x)y = ab.$$

The main reason why I believe this account by Heath is mathematically false, is that it obscures Archimedes' line of reasoning. Instead of allowing us to see why Archimedes' solution is valid, Heath's interpretation makes Archimedes' geometrical exposition appear forced and arbitrary. To Heath, the meaning of the problem is a certain equation, $x^2(a - x) = bc^2$, the meaning of the hyperbola is another equation, $x^2 = (c^2/a)y$, while the meaning of the parabola is yet a third equation, $(a - x)y = ab$. The problem gives rise to the hyperbola and the parabola, apparently just because equations give rise to equations, by algebraic manipulation (though, in fact, Heath does not try to trace the equations to their origins on such terms: no such purely algebraic derivation naturally compels us to derive the hyperbola and the parabola from the problem). But we can see that, for Archimedes, different meanings can be assigned to the objects involved. The problem means not an equation, but a certain configurational relation of lines and areas, which Archimedes makes concrete by embedding the problem within a pattern of parallel lines. The parabola then has the meaning – crucial to this particular problem – of simplifying a ratio involving areas to a ratio involving lines. In this way, the areas-and-lines proportion of the problem becomes a proportion involving four lines, all fitted within the pattern of parallel lines. In particular, it becomes a proportion involving lines along the diagonal of a rectangle. The meaning of the hyperbola, finally, is as an object that fixes lines along the diagonal of a rectangle. So it is in this way that mathematics has a history: objects have different meanings, according to the different practices to which they belong. Heath's conic sections are embedded within a practice of algebraic manipulations, and so their meaning is as a certain relation holding different variables together. Archimedes' conic sections are embedded within a practice of manipulations on geometric configurations, and so their meaning is as a tool for aligning objects within a configuration. Notice that Heath's and Archimedes' meanings do not relate to each other in any simple way (in particular, it is not the case that Archimedes' meanings are a subset of Heath's). When we move from Archimedes to Heath, some information is gained, while some is lost. Thus, it becomes impossible to recover Archimedes' solution within the sets of meanings available to Heath. At the introduction, we had suggested that a-historicism's best argument is that it allows us to recover the validity of a mathematical argument. But here we find an example where the promise of a-historicism – of gaining us a window into the past – is unfulfilled. It is precisely a historical reading that allows us to recover the validity of the ancient text.

We thus find – following the type of argument first suggested by Unguru (1975) – that geometrical algebra is wanting as a description of Greek mathematical practice. Having said that, we should point out, once again, the germs of the algebraic in Archimedes' formulation of the problem.

The fact is that algebraic reading is especially tempting in this case, because the geometrical configuration comes with ready-made orthogonal coordinates, namely the orthogonal asymptotes ΘΓΜ. This is part of a general orthogonal grid, within which the proof is conducted (for instance, as mentioned above, a feature of the argument is the identification between linear segments which lie directly above or below each other, through *Elements* 1.34).[43] Most importantly, the orthogonal grid is necessary to make the hyperbola (which conserves rectangles, rather than general parallelograms) conserve the equality between the areas ΘΛ = ΘΗ.

Within this orthogonal grid, the parabola, the hyperbola, and even, if you wish, the "cubic equation"

$$x^2(a - x) = bc^2$$

are all easily defined in terms of multiplications and subtractions involving what Heath calls "x" and "y."

It must be stressed immediately that this is an exceptional situation. Generally, Greek conic sections (and similar lines) are not embedded within a system of lines comparable to our modern coordinates. Rather, Greek conic sections appear floating in uncharted

[43] E.g., Steps 2, 3 of the analysis. For such an identification we do not need orthogonality, but more general parallelism. Still, the fact that everything is perpendicular is a useful simplification.

space. The proportions defining the sections are implicit, in the sense that they do not appear in the diagram. And when there is a configuration consisting of more than a single section, the proportions defining the two will generally be essentially unrelated. It is therefore a special feature of this diagram that the two asymptotes are perpendicular, that one of the asymptotes of the hyperbola is parallel to the axis of the parabola, while the parameter of the parabola lies on the other asymptote, which is in turn parallel to the line from which the solid magnitudes studied by the proposition are generated. So it becomes very plausible to think of ΘΓ as "the y axis," and of ΓM as "the x axis." Once again, modern possibilities seem to emerge. Why is that?

The most fundamental reason why the diagram of this proposition takes a grid-structure is, as mentioned above, that the hyperbola must be made to equate certain rectangles, which are simultaneously defined in terms of (i) a parabola, (ii) a system of parallel lines. ZH, the axis of the parabola, must be parallel to ΘΓ, which is one of the asymptotes of the hyperbola. ΘΓ in turn must be perpendicular to the other asymptote ΓM (for Apollonius' *Conics* II.12 to apply to the required areas). ΓM, finally, must be tangent to the parabola at H (for the property of the parabola to apply to the required lines). In short, the constructions of the parabola and the hyperbola are intertwined. In this way, once again, a certain inherent complexity of the problem serves – without Archimedes intending so – to suggest a further, non-geometric meaning of the objects involved. The parabola and the hyperbola each arise out of a specific configurational need. In the case of the parabola, this is to simplify areas into lines. In the case of the hyperbola, this is to align lines together along a diagonal. But because the constructions of the two conic sections are intertwined, they also become, incidentally, interrelated. Besides each serving its own specific geometrical function, they also happen to be defined relative to the same lines so that one can – if one wishes to – describe them as functionally interrelated. Heath did and, while Archimedes clearly did not intend any such functional understanding of the conic sections, it is interesting to see, once again, how the trace of a possible equation appears within Archimedes' geometrical problem.

Let us now recapitulate the three dualities we came across in Archimedes' treatment of the problem.

First, while the problem arises from a particular configuration involving spheres, Archimedes abstracts away from the sphere to treat a more general problem, involving relations between any lines and areas (Section 1.1).

Second, while Archimedes treats objects through strict geometrical operations, he also uses the strange phrase, "the {area} on the {line}" which suggests the operation of multiplying an area by a line (we shall return to discuss this duality in the next chapter).

Third, while Archimedes conceives of the conic sections as tools effective for producing relations within special geometrical configurations, he also happens to produce them in such a way that they can be defined in terms of a functional relation uniting them.

The three, put together, suggest the nature of the trajectory leading from problems to equations. Within geometrical problems, objects are considered as participating in local configurations, and they are manipulated to obtain relations within such a local configuration. In equations, objects belong to more general structures, and are related to each other in more general ways, independently of the local configurations they happen to be in.

Notice that we begin to conceive of the relation between problems and equations as a matter of *degree*. This, of course, is valuable for the historical account. But it has to be admitted: the account above is vague, and still a-historical. We need to specify further the nature of the difference between problems and equations, and we need to give a historical account: why did Hellenistic mathematicians produce problems, rather than equations? To answer this, we should accumulate more evidence. In the next two sections, we consider the two further known appearances of the same problem in Hellenistic mathematics.

1.4 The problem solved by Dionysodorus

There is nothing strange about Greek mathematicians solving problems that had already been solved by previous mathematicians. Indeed, it seems that much of the accumulation of techniques in Greek mathematics was the result of such competitive attempts at offering better solutions for already solved problems.[44] In this case, however, there was a special urgency: Archimedes, as far as Hellenistic mathematics was concerned, did not completely solve the problem. As we recall, in Proposition 4 of the *Second Book on the Sphere and Cylinder* Archimedes reduced the problem of cutting a sphere into the general problem of proportion of lines and areas. Following that, he went on as if the general problem was solved, and promised to solve it in an appendix to this *Book on the Sphere and Cylinder*. Whether or not such an appendix ever was attached to the main book in Antiquity, it clearly was lost from it at a very early stage, and it appears that most Hellenistic mathematicians did not have access to Archimedes' solution. It is only Eutocius, in his commentary to Archimedes – written in the sixth century AD – who claims to have found this Archimedean solution, and to reproduce it. (We shall return to discuss the trustworthiness of Eutocius' account, in the next chapter.)

The upshot of the situation was that, for an ancient mathematician, Archimedes' claim – to have solved the problem of cutting the sphere according to a given ratio – had been empty. Here, then, was an opportunity to better Archimedes himself. Dionysodorus and Diocles did so (apparently – in the second century BC). Let us first discuss Dionysodorus' solution, also reported to us by Eutocius (Heiberg [1915] 152.27–158.12):

To cut the given sphere by a plane, so that its segments will have to each other the given ratio.

Let there be the given sphere, whose diameter is AB, and <let> the given ratio be that which ΓΔ has to ΔE. So it is required to cut the sphere by a plane, right to AB, so that the segment whose vertex is A has to the segment whose vertex is B the ratio which ΓΔ has to ΔE.

(a) Let BA be produced to Z, (b) and let AZ be set <as> half of AB, (c) and let ZA have to AH <that> ratio which ΓE has to EΔ[45] (d) and let AH be at right <angles> to AB, (e) and let AΘ be taken <as> a mean proportional between ZA, AH; (1) therefore AΘ is greater than AH.[46] (f) And let a

[44] This process is the main theme of Knorr (1986).

[45] That is, the ratio ZA:AH transforms, into the concrete geometrical configuration of the sphere, the given, abstract ratio ΓE:EΔ. Notice that the diagram, somewhat confusingly, implies that ZA = AH, while the text demands ZA > AH. This is typical of the way in which Greek diagrams tend to ignore metrical considerations.

[46] AΘ is greater than AH, because it is the mean proportional in the series ZA – AΘ – AH (Step e), while ZA is greater than AH, because ZA, AH have the same ratio as ΓE, EΔ

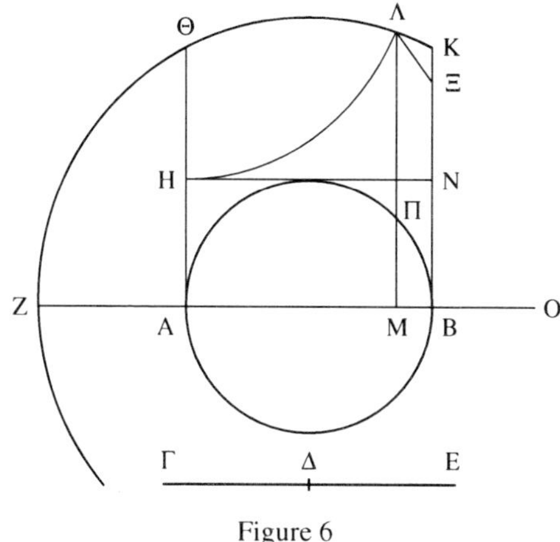

Figure 6

parabola be drawn through the <point> Z around the axis ZB, so that the <lines> drawn down <on the axis> are in square <the rectangles applied> along AH;[47] (2) therefore it shall pass through Θ, (3) since the <rectangle contained> by ZAH is equal to the <square> on AΘ.[48] (g) So let it be drawn, and let it be as the <line> ZΘK, (h) and let BK be drawn down through B, parallel to AΘ, (i) and let it cut the parabola at K, (j) and let a hyperbola be drawn through H, around ZBK <as> asymptotes; (4) so it cuts the parabola between the <points> Θ, K.[49] (k) Let it cut <the parabola> at Λ, (l) and let ΛM be drawn <as> a perpendicular from Λ on AB, (m) and let HN, ΛΞ be drawn through H, Λ parallel

(Step c), and ΓE is greater than EA – which, finally, we know from the position of the point Δ in the diagram. (Or, better still, we know this since the general Greek practice is to present ratios, when possible, in the order of the greater to the smaller.)

47 The meaning of this is that, for any point such as K on the parabola, $KB^2 = BZ*AH$. AH is what is known as the *latus rectum* of the parabola: in Archimedes' parabola, this was the line (in the analysis) HM. Notice that, here, the *latus rectum* is *not* at the vertex of the parabola.

48 Step e, *Elements* VI.17. Step 2 derives from Step 3 on the basis of the converse of *Conics* I.11.

49 The key insight of Dionysodorus' solution is that the hyperbola cuts the parabola at the relevant "box." This is stated without proof, typical for such topological insights in Greek mathematics (we shall see a similar case in the next chapter). Dionysodorus' understanding of the situation might have been like this. Concentrate on the wing of the hyperbola to the right of AΘ. It must get closer and closer to the line BK, without ever touching that line (BK is an asymptote to the hyperbola: the relevant proposition is *Conics* II.14). So the hyperbola cannot pass wholly below or above the point K; at some point, well before getting as far as the line BK, it must pass higher than the point K. Since at the stretch ΘK, the parabola's highest point is K (this can be shown directly from the construction of the parabola, *Conics* I.11), what we have shown is that the hyperbola, starting below the parabola (H below Θ), will become higher than the parabola, well before either gets as far as the line BK. Thus they must cut each other.

to AB. (5) Now since HΛ is a hyperbola, (6) and ABK are asymptotes, (7) and MΛΞ are parallel to AHN, (8) the <rectangle contained> by AHN is equal to the <rectangle contained> by MΛΞ, (9) through the 8th theorem of the second book of Apollonius' *Conic Elements*.[50] (10) But HN is equal to AB,[51] (11) while ΛΞ <is equal> to MB; (12) therefore the <rectangle contained> by ΛMB is equal to the <rectangle contained> by HAB, (13) and through the <fact> that the <rectangle contained> by the extremes is equal to the <rectangle contained> by the means, (14) the four lines are proportional;[52] (15) therefore it is: as ΛM to HA, so AB to BM; (16) therefore also: as the <square> on ΛM to the <square> on AH, so the <square> on AB to the <square> on BM. (17) And since (through the parabola), the <square> on ΛM is equal to the <rectangle contained> by ZM, AH,[53] (18) therefore it is: as ZM to MΛ, so MΛ to AH;[54] (19) therefore also: as the first to the third, so the <square> on the first to the <square> on the second and the <square> on the second to the <square> on the third;[55] (20) therefore as ZM to AH, so the <square> on ΛM to the <square> on HA. (21) But as the <square> on ΛM to the <square> on AH, so the <square> on AB was proved <to be> to the <square> on BM; (22) therefore also: as the <square> on AB to the <square> on BM, so ZM to AH. (23) But as the <square> on AB to the <square> on BM, so the circle whose radius is equal to AB to the circle whose radius is equal to BM;[56] (24) therefore also: as the circle whose radius is equal to AB to the circle whose radius is equal to BM, so ZM to AH; (25) therefore the cone having the circle whose radius is equal to AB <as> base, and AH <as> height, is equal to the cone having the circle whose radius is equal to BM <as> base, and ZM <as> height; (26) for such cones, whose bases are in reciprocal proportion to the heights, are equal.[57] (27) But the cone having the circle whose radius is equal to AB <as> base, and ZA <as> height, is to the cone having the same base, but <having> AH <as> height, as ZA to AH,[58] (28) that is ΓE to EΔ ((29) for, being on the same base, they are to each other as the heights);[59] (30) therefore the cone, too, having the circle whose radius is equal to AB <as> base, and ZA <as> height, is to the cone having the circle whose radius is equal to BM <as> base, and ZM <as> height, as ΓE to EΔ. (31) But the cone having the circle whose radius is equal to AB <as> base, and ZA <as> height, is equal to the sphere,[60] (32) while the cone having the circle whose radius is equal to BM <as> base, and ZM <as> height, is equal to the segment of the sphere, whose vertex is B, and <whose> height is BM, (33) as shall be proved further later on;[61] (34) therefore the sphere, too, has to the said segment the ratio which ΓE has to

[50] What we call *Conics* II.12. [51] *Elements* I.34. [52] *Elements* VI.16.
[53] *Conics* I.11. [54] *Elements* VI.17. [55] *Elements* VI.20 Cor. II.
[56] *Elements* XII.2. [57] *Elements* XII.15. [58] *Elements* XII.14.
[59] This belated explicit reference to *Elements* XII.14 is meant to support Step 27, not Step 28. It is probably Eutocius' contribution and, if so, so are probably the other references to the *Elements* and the *Conics*.
[60] *SC* I.34.
[61] Dionysodorus, at least as reported by Eutocius, does indeed proceed to offer this lemma. I do not reproduce it here, as it does not touch on our main theme.

EΔ; (35) dividedly, also: the segment, whose vertex is A, and <whose> height is AM, has to the segment, whose vertex is B, and <whose> height is BM, this ratio, which ΓΔ has to ΔE.[62] (36) Therefore the plane produced through ΛM, right to AB, cuts the sphere according to the given ratio; which it was required to do.

To follow Dionysodorus' line of reasoning, we should start from the problem as he had inherited it from Archimedes – that is, as stated for a particular setting arising from a sphere. Archimedes' statement was:

Therefore it is required to cut a given line, ΔZ, at the <point> X and to produce: as XZ to a given <line, namely> ZΘ, so the given <square, namely> the <square> on BΔ to the <square> on ΔX.

Which, adapted to Dionysodorus' labeling of the diagram, corresponds to:

Therefore it is required to cut a given line, BZ, at the <point> M and to produce: as MZ to a given <line . . . to be defined separately> so the given <square, namely> the <square> on AB to the <square> on BM.

Archimedes concretized the given ratio, by introducing the line ZΘ into his diagram of *SC* II.4, so that ΘZ:ΘB is the given ratio. Dionysodorus, who did not follow Archimedes' special route to the problem, concretized the ratio more directly, by introducing AH into his diagram, so that ZA:AH is the given ratio. A simple operation on ratios would show Dionysodorus that, with the new formulation of the ratio, what he requires is that the line MZ shall have the given ratio to the line AH. Hence the problem becomes, effectively, for Dionysodorus:

Therefore it is required to cut a given line, BZ, at the <point> M and to produce: as MZ to a given <line> AH[63] so the given <square, namely> the <square> on AB to the <square> on BM.

This proportion Dionysodorus obtains in Step 22 of his solution. It will be seen that, from Step 23 onwards, all Dionysodorus does is to show how, given this condition, the sphere had been cut

62 *Elements* V.17.

63 AH is given, since it is the consequent of a given ratio, whose antecedent is given (=the radius of the given sphere).

according to a given ratio. This is all the more striking, given that this relation had already been proved by Archimedes in his original discussion on cutting a sphere – a discussion that must have been still available to Dionysodorus himself. We begin to perceive a surprising mathematical practice. We shall return to discuss this when we consider the nature of Dionysodorus' solution – why it had approached the proposition as a problem, rather than as an equation.

For the time being, let us concentrate on Dionysodorus' solution up to and including Step 22. How did he reach, independently of Archimedes, the solution of the problem?

The only evidence we have for this is the proof itself: but, in close inspection, the proof is seen to follow a simple route, which is therefore very likely to have been Dionysodorus' own. We see Dionysodorus first producing the overall construction, in Steps a–m, 1–4. He then unpacks the consequences of the hyperbola, in Steps 5–16. Finally, he briefly unpacks the consequences of the parabola, in Steps 17–20. Finally, in Steps 21–2, the two separate consequences are merged to a single conclusion. It seems as if Dionysodorus thought primarily in terms of the hyperbola, as solving the problem, the parabola being an added tool for fixing the solution. (This is somewhat opposite to Archimedes' own approach.) Let us see how this may be done.

Recall the stated goal: MZ:AH::(sq. AB):(sq. BM). Now, we have a pattern with one straight line, ZO, on which we have two line segments, one within another: AB and BM. These two define one ratio in a proportion, (sq. AB):(sq. BM), while the other ratio contains a line, perpendicular to these two segments at one of their ends, A – the line AH. This immediately suggests thinking in terms of the basic relation that brings together plane geometry and proportion theory: *Elements* VI.16. This allows us to move from a proportion in four terms, to an equality between two rectangles, and vice versa. Thus, if we imagine the rectangles of the diagram erected on the given pattern – rectangles such as the rectangle ΞΛMB and the rectangle NHAB – then, if we arrange that the two rectangles are equal, we shall have a proportion involving the four sides of the rectangles: AB:MB::ΛM:AH or, if we wish, (sq. AB):(sq. MB)::(sq. ΛM):(sq. AH). Now, as we recall, arranging that two rectangles be equal is a simple task: all we need to do is to draw a hyperbola through the point H, with ΞB, BA as asymptotes. All are given by the terms of the problem, so that we can count on being able to have the proportion (sq. AB):(sq. MB)::(sq. ΛM):(sq. AH) when we need it.

Of course, this is, once again, a "sliding" configuration: we do not have a definite point M (this is what we seek to find in this problem). What we know is that, for any choice of a point M, we shall hit the hyperbola at a different point Λ – always conserving the proportion above. The question arises, then: how best to fix the point Λ?

Now, our goal is another proportion involving the ratio (sq. AB):(sq. MB). Our goal is the proportion MZ:AH::(sq. AB):(sq. BM). In other words, we wish to fix the point Λ in such a way, that MZ:AH::(sq. ΛM):(sq. AH). What is the geometrical significance of that? It is obvious (with the fundamental *Elements* VI.1) that, if we turn the first two lines, MZ and AH, into areas with the common height AH, their ratio would remain the same. In other words, we have (rect. MZ, AH):(sq. AH)::(sq. ΛM):(sq. AH) or, very simply:

rect. (MZ, AH) = sq. ΛM.

This immediately defines a parabola with Z as vertex, ZO as axis, and AH as *latus rectum*. Thus the problem is solved, once again, with the intersection of a hyperbola and a parabola. The hand of mathematical truth led both Archimedes and Dionysodorus to the same configuration of conic sections.

Mathematical truth, as we have seen above, may appear in different guises. To Archimedes, it was geometrical; to Heath, it was algebraic. Where, in this sense, stands Dionysodorus' proposition? Where is it in the trajectory leading from problems to equations?

The answer is interestingly complicated. In some ways, Dionysodorus' thought is more purely geometrical than Archimedes' while, in others, it is less geometrical and more purely quantitative.

Dionysodorus' conception is more geometrical, first of all, in the basic setting of the problem. For Archimedes, this is a general problem, involving any combination of lines and areas, and subsisting, in principle, independently of any particular geometrical configuration. Dionysodorus could easily have formulated his solution in such general terms. In fact, nothing in his proof relies on the specific properties of the configuration: for instance, no use is made, inside the solution itself, of the equality ZA = (radius of circle AΠB). Dionysodorus had chosen (always assuming Eutocius' report is trustworthy in this respect) to present his argument only for the case arising out of the cutting of the sphere. His diagram was made to include an inert circle, AΠB, which does not participate in the solution. The hyperbola and the parabola are arbitrarily made to relate to this space, with its implied sphere, circles, and cones. Thus they are less general than Archimedes' hyperbola and parabola. Dionysodorus' conic sections are tools for cutting a sphere; Archimedes' conic sections are tools for obtaining general proportions.

Having said that, it appears, from the way in which Dionysodorus makes his conic sections appear inside the proposition, that he conceives of them, in fact, in a more purely quantitative way than Archimedes did in his solution. Archimedes has introduced the hyperbola as a tool for aligning lines along a diagonal. Dionysodorus introduces the hyperbola as a tool for obtaining equality between rectangles, in turn understood as a tool for obtaining a proportion. For Archimedes, the hyperbola served to fix objects inside a configuration; for Dionysodorus, it serves for stipulating a relation between four lines, abstracting away from their configuration. Archimedes introduced the parabola as a tool for reducing statements about abstract areas, into statements about tangible lines, participating in the configuration. Dionysodorus introduces the parabola because it is defined by an abstract relation that arises directly from proportion-manipulations. For Archimedes, the parabola served to transform a problem into terms of linear configuration; for Dionysodorus, it serves as a geometrical representation of a purely quantitative relation.

The apparent paradox – that Dionysodorus' basic setting is more geometrical, while his approach in the solution itself is more abstract – is in fact easy to account for. It is just because Archimedes had moved to a separate diagram, abstracting away from the sphere, that he could force the configuration as he wished to. Thus, he created from scratch a geometrical correlate for the proportion statements, in the pattern of similar triangles inside a rectangle. Everything else in the argument followed from this basic geometrical embedding of proportion statements. Dionysodorus, on the other hand, keeping to the configuration of the problem of the sphere, did not have a similar geometrical embedding of the proportion relations. He therefore treated those proportions in a more abstract way, as purely quantitative relations.

We should once again note, then, a certain duality. We begin to perceive a dialectical relation between the "geometrical" and the "abstract" (which we may even refer to as the "algebraic"). The two do not rule out each other: they coexist in complicated ways. Having said that, however, it is clear that Dionysodorus' argument, while relying on a more quantitative understanding of its objects, does not solve an equation. Dionysodorus might be relying upon equation-type bits of information (AB:MB::ΛM:AH for the rectangles, rect. (MZ, AH) = sq. ΛM for the parabola), but he simply nowhere presents an equation to be solved. To the reader unfamiliar with Archimedes' general statement of the problem, the proportion at Dionysodorus' Step 22 appears as a mere step along the way, and not as the goal of the argument. Thus we have learned two things: that the trajectory, from problems to equations, is complicated and many-dimensional; and that, with Dionysodorus, we are still essentially within the world of problems.

Why did Dionysodorus move away from Archimedes' position at all? After all, given that he had not available to him Archimedes' own proof, it is even conceivable that he, Dionysodorus, could have hit, by chance, Archimedes' own solution! Let us try and see why he did not.

The crux of the difference between Archimedes' and Dionysodorus' solutions lies, as we had seen, in Dionysodorus' decision to solve the problem in the particular terms of the cutting of the sphere. Now, in mathematical terms, it made little sense for Dionysodorus to do so. We should perhaps look for extra-mathematical reasons: and an obvious one may be offered along the following lines.

For Dionysodorus to solve the general problem would be to admit that Archimedes had effectively already solved the problem of cutting the sphere, merely assuming the lemma of the general problem of a proportion of lines and areas. If you wish to cut a sphere, read Archimedes and learn how to do so. If you wish to complete this in full rigor, you may then also consult Dionysodorus. In other words, Dionysodorus appears as a mere footnote to Archimedes. On the other hand, by fixing the problem as that of cutting a sphere, Dionysodorus suggests the following: that he is the first to offer a solution to a problem that Archimedes had only *claimed* to have solved. Dionysodorus appears then as an equal to Archimedes: indeed, in this particular case, he goes one better than Archimedes himself. That Dionysodorus wished to appear in this way is supported, I would argue, by his conclusion of the argument, from Step 23 onwards. As mentioned above, Dionysodorus, in effect, recapitulates Archimedes' argument as available to him, *presenting it as his own*. It was this, essentially polemical stance, that, I suggest, led Dionysodorus on his own route to the problem, approaching it directly in terms of the cutting of the sphere. In order to single himself out from Archimedes, he was forced to approach the problem not in any terms, but in the precise terms from which it arises in Archimedes' original problem.

What may that teach us? Now, it is important to stress, one cannot generalize and say that Archimedes is always a more "geometrical" author. He does resort in his writings to more abstract manipulations of quantities, moving away from concrete geometrical configurations (this is most often done in one of his most advanced works, *On Conoids and Spheroids*). As for Dionysodorus, we cannot of course make any general statements about him as the above is, in fact, the only direct evidence we have for him as a mathematician. Thus, it would be a mistake to make any general claims about Archimedes and Dionysodorus, say, that "already in Dionysodorus, we see a movement towards a more algebraic conception of the conic sections." This is clearly unwarranted. However, we do begin to see a certain dynamics at work. Because Dionysodorus has to face a problem that is already given under specified conditions and configurations, he has less freedom than Archimedes had. Since he cannot flex the configuration, he must flex the tools with which he approaches the configuration. Hence his conic sections appear less natural, more purely quantitative, than Archimedes' did. This may remind us of Klein's thesis: by being dependent on a given past, mathematics becomes more second-order, more abstract. With Dionysodorus, we see no more than a suggestion of this dynamics: we shall see much more of it in the next chapters.

1.5 The problem solved by Diocles

Following his excerpt from Dionysodorus, Eutocius went on to quote yet another solution to the problem of cutting the sphere, this time by Diocles. In this case, Eutocius' text is further corroborated by an Arabic translation of Diocles' original treatise (Toomer [1976], Rashed [2000]). Furthermore, we have a better sense of Diocles' project: we have ascribed to him both an analysis and a synthesis and, what is most important, a general introduction to his treatment of the problem.

We do not know the chronological relation between Dionysodorus and Diocles. Neither refers to the other. Perhaps Diocles came first, and this is why he did not mention Dionysodorus (Dionysodorus' introduction – if there was one – is not preserved). But I do not believe such conclusions are warranted. Whether or not Diocles had access to Dionysodorus' solution, he had no interest in referring to it. Here was an opportunity to challenge Archimedes: a mention of Dionysodorus would not make the challenge any more impressive. We can say, then, that, regardless of the chronological details, Diocles and Dionysodorus were *mathematical* contemporaries: both worked after Archimedes, and in direct reaction to him. Let us proceed to read Diocles, then, to see if the similar historical context led to a similar mathematical approach. The text translated here is from Eutocius' Greek version, Heiberg (1915) 160.3–168.25:

As Diocles in *On Burning Mirrors*[64]

And Diocles, too, gives a proof, following this introduction:

Archimedes proved in *On Sphere and Cylinder* that every segment of a sphere is equal to a cone having the same base as the segment, and, <as> height, a

[64] The following text also corresponds (very closely, though not exactly) to Propositions 7–8 of the Arabic translation of Diocles' treatise (Toomer [1976] 76–86, Rashed [2000] 119–25. Toomer also offers at 178–92 a translation of the passage in Eutocius with a very valuable discussion, 209–12). We shall return to the divergences between the Greek and the Arabic below.

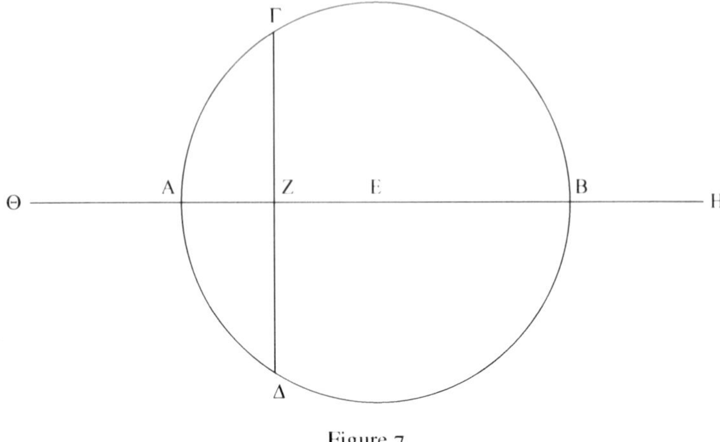

Figure 7

line having a certain ratio to the perpendicular <drawn> from the vertex of the segment on the base: <namely, the ratio> that: the radius of the sphere, and the perpendicular of the alternate segment, taken together, have to the perpendicular of the alternate segment.[65] For instance, if there is a sphere ABΓ, and if it is cut by a certain plane, <namely> the circle around the diameter ΓΔ,[66] and if (AB being diameter, and E center) we make: as EA, ZA taken together to ZA, so HZ to ZB, and yet again, as EB, BZ taken together to ZB, so ΘZ to ZA, it is proven: that the segment of the sphere ΓBΔ is equal to the cone whose base is the circle around the diameter ΓΔ, while its height is ZH, and that the segment ΓAΔ is equal to the cone whose base is the same, while its height is ΘZ. So he set himself the task of cutting the given sphere by a plane, so that the segments of the sphere have to each other the given ratio, and, making the construction above, he says: "(1) Therefore the ratio of the cone whose base is the circle around the diameter ΓΔ, and whose height is ZΘ, to the cone whose base is the same, while its height is ZH, is given, too;"[67] (2) and indeed, this too was proved;[68] (3) and cones which are on equal bases are to each other as the heights;[69] (4) therefore the ratio of ΘZ to ZH is given. (5) And since it is: as ΘZ to ZA, so EBZ taken together to ZB, (6) dividedly: as ΘA to AZ, so EB to ZB.[70] (7) And so through the same <arguments> also: as HB to ZB, so the same line <=EB> to ZA.

So a problem arises like this: with a line, <namely> AB, given in position, and given two points A, B, and given EB, to cut AB at Z and to add ΘA, BH so

[65] *SC* II.2.

[66] The circle meant is that perpendicular to the "plane of the page," or to the line AB.

[67] This text is part Diocles' own analysis, part a recreation of Archimedes' analysis, now in the terms of Diocles' diagram. Step 1 here corresponds to *SC* II.4 Step 4.

[68] Step 2 probably means: "by proving *SC* II.2, we thereby prove the claim of Step 1."

[69] *Elements* XII.14. [70] *Elements* V.17.

that the ratio of ΘZ to ZH will be <the> given, and also, so that it will be: as ΘA to AZ, so the given line to ZB, while as HB to BZ, so the same given line to ZA.

And this is proved in what follows. For Archimedes, having proved the same thing, rather long-windedly, even so he then reduced it to another problem, which he does not prove in the *Sphere and Cylinder*!

Given in position a line AB, and given two points A, B, and the ratio, which Γ has to Δ, to cut AB at E and to add ZA, HB, so that it is: as Γ to Δ, so ZE to EH; and also that it is: as ZA to AE, so a certain given line to BE, and as HB to BE, so the same given line to EA.[71]

(a) Let it come to be, (b) and let ΘAK, ΛBM be drawn at right <angles> to AB, (c) and let each of AK, BM be set equal to the given line. (d) Joining the <lines> KE, ME, let them be produced to Λ, Θ, (e) and let KM be joined, as well, (f) and let ΛN be drawn through Λ parallel to AB, (g) and let ΞEOΠ <be drawn> through E, <parallel> to NK. (1) Now since it is: as ZA to AE, so MB to BE; (2) for this is assumed; (3) and as MB to BE, so ΘA to AE (4) through the similarity of the triangles,[72] (5) therefore as ZA to AE, so ΘA to AE. (6) Therefore ZA is equal to ΘA.[73] (7) So, through the same <arguments>, BH, too, <is equal> to BΛ.[74] (8) And since it is: as ΘAE taken together to MBE taken together, so KAE taken together to ΛBE taken together; (9) for each of the ratios is the same as the <ratio> of AE to EB;[75] (10) therefore the <rectangle contained> by ΘAE taken together and by ΛBE taken together, is equal to the <rectangle contained> by KAE taken together and by MBE taken together;[76] (h) Let each of AP, BΣ be set equal to KA.[77] (11) Now since ΘAE taken together is equal to ZE, (12) while ΛBE taken together is equal to EH, (13) and KAE taken together is equal to PE, (14) and MBE taken together is equal to ΣE, (15) and the <rectangle contained> by ΘAE taken together and by ΛBE taken together was proved to be equal to the <rectangle contained> by KAE taken together and by MBE taken together, (16) therefore the <rectangle contained> by ZEH

[71] The "certain given line" remains unlabeled.

[72] The triangles referred to are ΘAE, BEM. That they are similar can be seen through Step b, *Elements* I.27, I.29, I.15 (or I.29, I.32). Step 3 derives from Step 4 through *Elements* VI.4.

[73] *Elements* V.9.

[74] The setting-out and Step a, again, provide the proportion HB:BE::KA:AE and, through the similarity of the triangles KAE, ΛEB the argument is obvious.

[75] By "each of the ratios" Diocles refers to the ratios of the separate lines making up the "taken together" objects. So we have four ratios: ΘA:MB, AE:BE, KA:ΛB, AE:BE (AE:BE occurs twice). All, indeed, are the same as AE:BE, through the similarities of triangles we have already seen. Step 8 follows from Step 9 through successive applications of *Elements* V.12.

[76] *Elements* VI.16. Notice a possible source of confusion. The rectangles are each contained by two lines, and each of these lines is a sum of two lines, denoted by three characters. This is confusing, because often we have a rectangle contained by two lines, and these containing two lines are directly denoted by three characters. Here the summation happens not between the sides of the rectangles, but inside each of the sides.

[77] Thus all lines AP, BΣ, KA, BM are now equal to the unlabeled, given line – this anonymous line is cloned, as it were, all through the diagram.

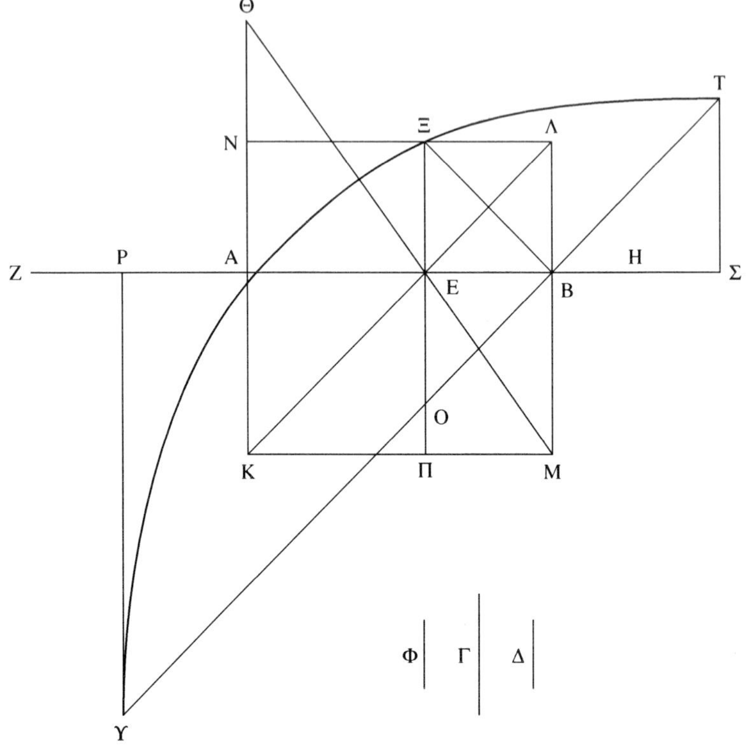

Figure 8

is equal to the <rectangle contained> by PEΣ. (17) So through this, whenever P falls between the <points> A, Z, then Σ falls outside H, and vice versa.[78] (18) Now since it is: as Γ to Δ, so ZE to EH, (19) and as ZE to EH, so the <rectangle contained> by ZEH to the <square> on EH,[79] (20) therefore: as Γ to Δ, so the <rectangle contained> by ZEH to the <square> on EH. (21) And the

[78] The "vice versa" means that, conversely to what has been mentioned, also when Σ falls between B, H, then P falls outside Z. ("Outside" here means "away from the center of the diagram" – imagine the diagram as an underground network, and imagine that the lines have two directions, "Inbound" and "Outbound.") This is a remarkable moment. The author of this passage is aware both of topological considerations, and of a functional relation between variables. Yet the underlying argument is very simple: it is impossible to have two equal rectangles, if the sides of one of the rectangles are both greater than the sides of the other. If one side is greater, the other must be smaller. This is not stated in the *Elements*, but it is implicit in *Elements* VI.16. (That P, Σ, must both be "outside" AB, is implicit in the construction of the points and is, in practice, learned from the diagram).

[79] *Elements* VI.1.

<rectangle contained> by ZEH was proved equal to the <rectangle contained> by PEΣ; (22) therefore it is: as Γ to Δ, so the <rectangle contained> by PEΣ to the <square> on EH. (i) Let EO be set equal to BE, (j) and, joining BO, let it be produced to either side, (k) and, drawing ΣT, Pϒ from Σ, P at right <angles to the line AB>, (l) let them meet it <=the line BO, produced> at T, ϒ. (23) Now since the <line> Tϒ has been drawn through a given <point> B, (24) producing, to a <line> given in position, <namely> to AB, an angle (<namely>, the <angle contained> by EBO), half of a right <angle>,[80] (25) Tϒ is given in position.[81] (26) And the <lines> ΣT, Pϒ, <given> in position, drawn from given <points,> Σ, P, cutting it <=the line Tϒ, given in position,> at T, ϒ; (27) therefore T, ϒ are given;[82] (28) therefore Tϒ is given in position and in magnitude. (29) And since, through the similarity of the triangles EOB, ΣTB,[83] (30) it is: as TB to BO, so ΣB to BE,[84] (31) it is compoundly, also: as TO to OB, so ΣE to EB.[85] (32) But as BO to Oϒ, so BE to EP.[86] (33) Therefore also, through the equality: as TO to Oϒ, so ΣE to EP.[87] (34) But as TO to Oϒ, so the <rectangle contained> by TOϒ to the <square> on Oϒ, (35) and as ΣE to EP, so the <rectangle contained> by ΣEP to the <square> on EP;[88] (36) therefore also: as the <rectangle contained> by TOϒ to the <square> on Oϒ, so the <rectangle contained> by ΣEP to the <square> on EP; (37) alternately also: as the <rectangle contained> by TOϒ to the <rectangle contained> by ΣEP, so the <square> on Oϒ to the <square> on EP. (38) And the <square> on Oϒ is twice the <square> on EP, (39) since the <square> on OB is twice the <square> on BE, too;[89] (40) therefore the <rectangle contained> by TOϒ, too, is twice the <rectangle contained> by ΣEP. (41) And the <rectangle contained> by ΣEP was proved to have, to the <square> on EH, the ratio which Γ has to Δ; (42) and therefore the <rectangle contained> by TOϒ has to the <square> on EH the ratio, which twice Γ has to Δ. (43) And the <square> on EH is equal to the <square> on ΞO; (44) for each of the <lines> EH, ΞO is equal to ΛBE taken together;[90] (45) therefore the <rectangle contained> by TOϒ has to the <square> on ΞO <the> ratio, which twice Γ has to Δ. (46) And the ratio of twice Γ to Δ is given; (47) therefore the ratio of the <rectangle contained> by TOϒ to the <square> on ΞO is given as well.

(48) Therefore if we make: as Δ to twice Γ, so Tϒ to some other <line>, as Φ, and if we draw an ellipse around Tϒ, so that the <lines> drawn down <on the diameter>, inside the angle ΞOB (that is <inside> half a right <angle>), are in square the <rectangles applied> along Φ, falling short by a <figure> similar

[80] From Step i, OE = EB. From Steps b, g, OEB is a right angle. Then the claim of Step 24 is seen through *Elements* I.32.

[81] *Data* 30. [82] *Data* 25. [83] Steps b, g, k, *Elements* I.27, 29, 15 (or 32).

[84] *Elements* VI.4. [85] *Elements* V.18. [86] Steps b, g, k, *Elements* I.27, VI.2.

[87] *Elements* V.22. [88] Steps 34–5: both from *Elements* VI.1.

[89] This is through the special case of Pythagoras' theorem (*Elements* I.47) for an isosceles right-angled triangle.

[90] ΞE = ΛB (through Steps b, f, g, *Elements* I.27, 30, 34). EO = EB through Step i. So this settles ΞO = ΛBE. EH = ΛBE can be seen through Step 7.

to the <rectangle contained> by TΥ, Φ,[91] <the ellipse> shall pass through the point Ξ, (49) through the converse of the twentieth theorem of the first book of Apollonius' *Conic Elements*.[92] (m) Let it be drawn and let it be as ΥΞT; (50) therefore the point Ξ touches an ellipse given in position. (51) And since ΛK is a diagonal of the parallelogram NM,[93] (52) the <rectangle contained> by NΞΠ is equal to the <rectangle contained> by ABM.[94] (53) Therefore if we draw a hyperbola through the <point> B, around ΘKM <as> asymptotes, it shall pass through Ξ,[95] (54) and it shall be given in position ((55) through the <facts> that the point B, too, is given in position, (56) as well as each of the <lines> AB, BM, (57) and also, through this, the asymptotes ΘKM.) (n) Let it be drawn and let it be as ΞB; (58) therefore the point Ξ touches a hyperbola given in position. (59) And it also touched an ellipse given in position; (60) therefore Ξ is given.[96] (61) And ΞE is a perpendicular <drawn> from it; (62) therefore E is given.[97] (63) And since it is: as MB to BE, so ZA to AE, (64) and AE is given, (65) therefore AZ is given, as well.[98] (66) So, through the same <arguments>, HB is given as well.[99]

At this point Eutocius went on to produce the synthetic solution: this was in all likelihood Eutocius' own contribution (since, in the Arabic version, Diocles explicitly ignores the synthesis as

[91] This is the Apollonian way of stating that Φ is the parameter of the ellipse. Imagine that Φ is set at the point T, at right angles to the line ΥT. You get a configuration similar to that of *Conics* I.13 (see fig. 7, Φ here is transformed into EΘ), for which Apollonius proves that for any point Λ taken on the ellipse EΛΔ, the square on ΛM is equal to the associated rectangle MO.

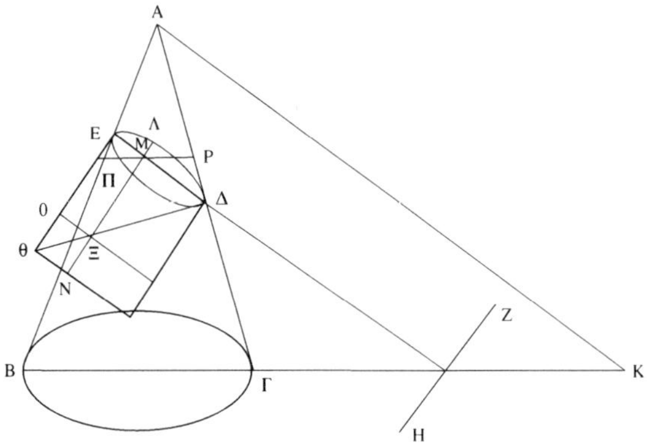

[92] What we call *Conics* I.21.

[93] Steps b, c, f, *Elements* I.27, 33.

[94] Based on *Elements* I.43.

[95] Converse of *Conics* II.12.

[96] *Data* 25.

[97] *Data* 30.

[98] With E given, BE is given as well. BM is given from setting-out, Step c, hence BM:BE is given. Step 65 then derives from *Data* 2.

[99] The only difference will be that instead of using the proportion MB:BE::ZA:AE, we use the proportion HB:BE::BM:EA (both from setting-out, Step c).

trivial).[100] This is then an example of the way in which Eutocius had transformed the materials available to him from the Hellenistic past: we shall return to see more examples of this in the next chapter.

The first thing to note for Diocles himself is that he went even further than Dionysodorus did, in retracing Archimedes' path. The difference is this. Dionysodorus went back to the problem in its particular form of cutting the sphere, adopting, effectively, Archimedes' analysis of the conditions for solving that problem. Dionysodorus merely pretended to do without Archimedes' analysis while, in fact, following Archimedes' statement. Thus Dionysodorus had merely pretended to solve the problem of cutting the sphere, and offered, in fact, a solution for the general problem of a proportion with lines and areas. Diocles went further back than Dionysodorus did, ignoring Archimedes' final statement of the condition. Diocles kept close to Archimedes' original setting of the conditions in terms of the sphere, and summed up the problem in those, as it were unprocessed terms. His statement considers the problem, then, not in terms of *a single proportion of lines and areas*, but in terms of *three proportions of lines*. To recall: Archimedes' analysis had led him to the following condition (in terms of fig. 2):

> To cut a given line, ΔZ, at the <point> X and to produce:
> XZ:ZΘ::(square BΔ): (sq. on ΔX). [The point Θ is indirectly defined in terms of the given ratio.]

Diocles' condition is, adapted to the same diagram (the translation is, from Archimedes' diagram to Diocles': Λ→H, Δ→B, K→E, X→Z, B→A, P→Θ):

> to cut ΔB at X and to add PB, ΛΔ so that:
> (i) PX:XΛ::(given ratio),
> (ii) PB:BX::(the given line <=ΔB>):XΔ,
> (iii) ΛΔ:ΔX::(the same given line <=ΔB>):XB.

Diocles' terms keep close to the meaning of the proposition as a problem of cutting the sphere. They also have the advantage that they do not involve areas and lines in the same proportion, but are all

[100] Rashed (2000) 125.16.

stated in terms of four-line proportions – the simplest stuff of Greek proportion theory. The obvious weakness of Diocles' statement is that it is triple: Diocles needs to do three things simultaneously (*cut a line*, and *add two lines*), obtaining three separate proportions. One must admire the beauty of Archimedes' analysis that simplifies the conditions into a single act (*cut a line*) and a single proportion. (The price paid for this, of course, is that Archimedes' proportion involves both areas and lines.) Why would Diocles prefer, then, the complicated three-task problem? The reason, once again, seems to be extra-mathematical. In this case, after all, we do have Diocles' introductory words which, explicitly, stress Diocles' desire to set himself apart from, and to criticize, Archimedes. For Diocles, Archimedes is to be criticized for reducing the problem to another, and then (apparently) failing to solve the other problem. Diocles would do better than Archimedes did, by keeping much closer to the original problem.

Further, Diocles takes a leaf of Archimedes' book, and conjures the problem – without ever saying so explicitly – into a more general plane. We saw that Diocles' conditions (i) and (ii) refer to a given line, by which Diocles means the radius of the sphere. However, Diocles' statement fails to mention that this given line happens to be this radius, and he proceeds to solve the problem assuming that this is indeed any given line. This is a very subtle move, compared to Archimedes' complete generalization of the problem, but it is significant: by treating the given line in general terms, the problem is already separate from that of cutting the sphere, while also being distinct from *Archimedes'* general problem. Once again, the motivation does not seem to be generalization in itself: Diocles, as we see, does not at all highlight this act of generalization as such. Instead, the operating urge seems to be to distinguish the problem, as sharply as possible, from Archimedes'. By moving away to a generalization, which however is different from Archimedes', Diocles protects his solution from being equated with Archimedes'. He solves neither the particular case of the cutting of the sphere, nor the general problem of a proportion with lines and areas. Instead, he solves another general problem, with three proportions of lines: a problem which happens to include that of cutting a sphere, but is stated in terms that are more general.

All of this sets Diocles a difficult task. How was he going to solve his three-task, three-proportion problem? Let us tentatively try and follow his possible line of thought. Our evidence is limited. The analysis is all we have from Diocles – a good example, then, of how indirectly the analysis is related to the process of discovery. For clearly the analysis, as it is set out in Diocles' text, came only *after* the discovery. Or else, how did he get his complicated construction, going so much beyond the terms of the proposition? We need to see how the construction could have been suggested by those terms. In fact, I believe this can be followed in detail. Be warned: the route ahead is long and winding.

Let us first rephrase the problem, now in terms of the diagram of Diocles' solution.

Given: AB, Γ:Δ, a certain given line = AK.
Task: to cut AB at E and to add ZA, HB, fulfilling:

(i) Γ:Δ::ZE:EH;
(ii) ZA:AE::AK:BE;
(iii) HB:BE::AK:EA.

We see that, merely in order to get us thinking, it is useful to put the general given line (the generalization of the radius) in some concrete form in the diagram and, just as Dionysodorus did, this line is put on the main given line, perpendicular to it.

Conditions (ii) and (iii) are intriguing: the ratio of a line to one line segment is the ratio of another line (perpendicular to the first line) to the line segment complementary to the first. There is a certain reciprocity within each of the two conditions – in the way in which the first consequent is, somehow, complementary to the second. Now, in general, when thinking about proportion, the first thing that comes to mind is similar triangles and, when we think of a reciprocity inside a proportion, this is very suggestive of similar triangles pivoting around a common point. Let us then try to concretize the conditions in terms of similar triangles pivoting around a common point. Take for instance condition (iii). AE and EB are the consequents: let us treat them as two bases of similar triangles. AK should stand in the given ratio to EA. Now extend the lines EΛ, ΛB to complete the similar triangles, and condition (iii),

HB:BE::AK:EA has a new significance. From the similar triangles we have ΛB:BE::AK:AE so that we now have the equality HB = ΛB. Sliding the line AK sideways, to obtain the position BM, and completing another couple of similar triangles, we obtain similarly the equality ZA = ΘA. Notice that the diagram could make us think of the central structure as primarily a rectangle, with an extended point Θ: this is misleading (much later on we will indeed need to see a rectangle here, but this will come only when we introduce the hyperbola). The central structure is primarily a set of four similar triangles along the points K–Θ–E–M–Λ, with the property that AK = BM. The points K, M are given by the terms of the problem, and, as soon as we have the extra point E (required by the first task of the problem) we immediately determine the remaining points Θ, Λ.

What can we do with this configuration and with this set of equalities, HB = ΛB, ZA = ΘA? Well, so far we have considered conditions (ii) and (iii). Let us see what more we can do with condition (i). This effectively determines the ratio ZE:EA which we can now, with the new equalities, re-identify as the ratio ΘAE:ΛBE (by which we mean ΘA + AE, ΛB + BE). Of course, everything in similar triangles is "made to scale" so that the ratio of a sum of lines is like the ratio of the lines themselves: this is helpful. But unfortunately, there is little we can do with the direct ratio between these two upper triangles, ΘAE, ΛBE. If we had one triangle from above the line, and another from below the line, we could directly transform the ratio. From the ratio ΘAE:EBM, for instance, we could directly obtain the ratio KAE:EBΛ. But this is not what we are given. The two upper triangles are to each other not as the two antecedents in a proportion, but as the first antecedent and the second consequent. In general, when we have the structure A:B::C:D, we can do all sorts of things with a ratio-couple such as A:C (or B:D, or indeed A:B, C:D); but the ratio-couple such as A:D is unhelpful. We just do not know anything about it. Such a couple as A and D (or B and C) is useful to us not as a ratio-couple, but as a rectangle-couple. That is, if we have the rectangle whose sides are A and D, we can immediately equate it with the rectangle whose sides are B and C. So there is after all something we can do with the given pair, ΘAE, ΛBE: we can manipulate the rectangle this pair defines. This rectangle is equal to the other rectangle defined by the central structure of similar triangles: (rect. ΘAE, ΛBE) = (rect. KAE, EBM). We can only use the pair as a rectangle-pair? No worry: we might as well consider such rectangles. Condition (i) is stated in the terms of a *ratio between lines*, ZE:EH being fixed. But turn both lines into areas with the common height EH, and you have a fixed ratio between *a rectangle* and a square: (rect. ZE, EH):(sq. EH). Rectangles matter to the problem: they are implied by condition (i). Now we return to the central structure of similar triangles and put it to good use. We know, effectively, that (rect. ZE, EH) has a fixed ratio to (sq. EH). In other words, we know that (rect. ΘAE, ΛBE) has a fixed ratio to (sq. EH). And we also know, thanks to the central structure of similar triangles, that (rect. ΘAE, ΛBE) = (rect. KAE, EBM). Now we have something remarkable, coming out of the three conditions: (rect. KAE, EBM) has a fixed ratio to (sq. EH).

At this point, we can proceed in a very elegant manner. Having folded conditions (ii) and (iii) into the central structure of similar triangles, we can now unfold them, since both lengths KAE, EBM are fully given in terms of E. KA, BM are not unknown lengths: they are among the given terms of the problem. So we can set them out along the basic line, with PA = AK, ΣB = BM. Now we have the relation: (rect. PE, EΣ) has a fixed ratio to (sq. EH).

Notice further that both rectangle and square pivot around a single point, E: the rectangle has its two sides on the two sides of E; the square has its side on a line ending at E. "A fixed ratio between a rectangle and a square, both pivoting at a single point" – this immediately defines an ellipse with the pivoting point as its intersection of its axis and one of the ordinates, and the fixed ratio defining its metrical property. To make this ellipse concrete, if we allow the line EH to extend perpendicularly (for simplicity's sake) from the point E – which we do not have in the diagram, but which we can imagine for the sake of the argument (as in fig. 9) – then we have an ellipse passing through the points P, H', Σ. It is defined by the fixed ratio (rect. ΣE, EP) has to (sq. EH), and – once we have that ellipse – it will solve the problem.

Most important, the lines PE, EΣ are both defined in terms of the point E so that, once we have the point E, we also have (rect.

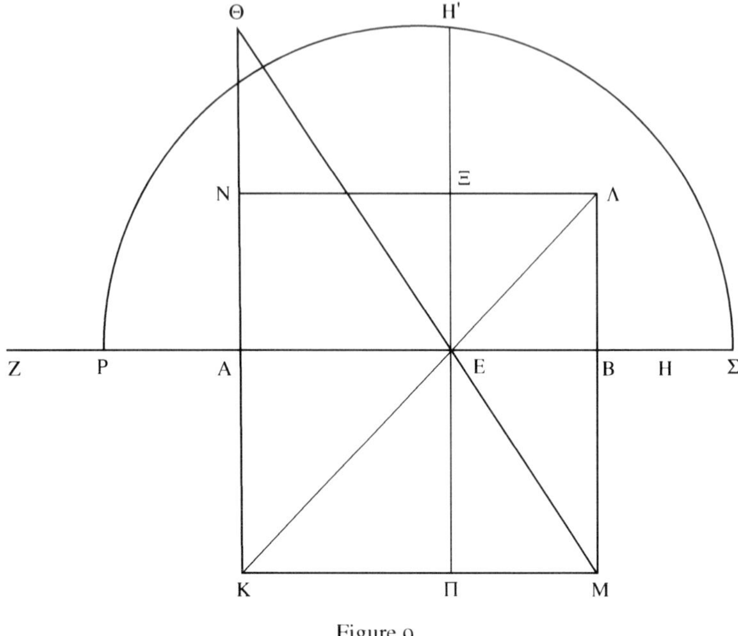

Figure 9

PE, EΣ). And since the fixed ratio is in itself known, once we know (rect. PE, EΣ) we also know (sq. EH) or (sq. EH'). Everything has been reduced to a single task: finding a point E, so that it yields a given ratio between a resulting rectangle and a resulting square or, in other words, so that it determines a definite ellipse.

If only we knew how to fix this ellipse! For, after all, we still do not have the point E. For any point E taken along the line AB, we have a different (rect. PE, EΣ), with a resulting different (sq. EH') and a different ellipse.

We therefore look for something extra we might know about the point E, to fix the ellipse in place. And we do in fact know something extra about E: assuming the problem has been solved, then this must be that point of AB, which is also found on the straight line KΛ. (This point is not yet settled just by the given terms of the problem: the position of Λ is defined by the ratio AK:EA, which we must find as condition (iii)). To put it differently: if we enclose finally the points K, Λ within the rectangle ΛNKM, then the point E must fall on the diagonal of the rectangle. We already know how to solve this: we reach our tool of aligning points along a line – the hyperbola – and draw a hyperbola passing through B with the two asymptotes NK, KM.

Is the problem solved now? Not yet. The hyperbola finds the diagonal by meeting the rectangle ΛNKM, that is by meeting the line ΛN – which still depends on the as yet unfound point Λ! In other words, in order to solve the problem, we need to ensure one final extra requirement: that the ellipse and the hyperbola shall somehow oblige us, and intersect each other exactly on the rectangle, that is, exactly so that ΞE = ΛB. And here comes the last brilliant observation: we do know something about this length ΛB, after all! It happens to satisfy the equality ΛBE = EH, that is ΛB = BH. With the ordinate EH' satisfying EH' = EH, we even have a tight relation between the length of the line ΛB, and the ellipse: ΛBE = EH', that is, the ordinate EH' projects out from the rectangle at a length exactly equal to EB. So, while we do not know where the point Λ is – that is, we do not know the precise length of the line ΞE – we can still ensure that the ellipse would exactly reach the point Ξ. We can simply allow the ordinate EH' to slide down below the line ZΣ, until its top exactly touches the rectangle. To do this, it should slide down by a length equal to EB – the original difference in length between EH' and ΛB. That is: when the bottom point of the ordinate has moved down from the point E to its new point O, we also have EO = EB, and the triangle BEO happens to be a very simple triangle – right-angled and isosceles. So finally, we can allow the ellipse PH'Σ to slide down, always keeping the length of its ordinate intact, its axis rotating by half a right angle. When it does so, it reaches the condition that the top of the ordinate reaches the line ΛN at the point Ξ. Its metrical property is of course somewhat changed, but this is merely by a fixed quantity: the square on OΞ, the new ordinate, is exactly the same as the square on EH', the old ordinate, since the two lines are equal, while (rect. TO, OY), the new rectangle on the segments of the axis, has been simply doubled. The ellipse YΞT is different from the ellipse PH'Σ, but it is just as easy to construct. And it has to solve the problem: the point where it cuts the hyperbola BΞ must define a rectangle at whose diagonal lies the point E, and at whose vertex lies the point Λ.

Whoof! That was a long and complicated argument: some arguments are. And I am reasonably confident that it does correspond to the main thought of Diocles' solution. The argument appears long, because it takes many turns. Yet each of the turns are largely dictated by their immediate background. I sum it up briefly. Conditions (ii) and (iii) are obviously useful, and by their reciprocity immediately suggest the idea of similar triangles on which to project the proportions they define. Thus arises the central scheme of similar triangles. This scheme gives rise to the first real discovery of the solution, HB = ΛB, ZA = ΘA. We then try to force this discovery – and this central scheme of similar triangles – to get us something out of condition (i), with its fixed ratio ZE:EH. This practically forces us into the second discovery of the solution, that (rect. KAE, EBM) has a fixed ratio to (sq. EH). It is very natural now to unfold KAE, EBM so as to get, instead, the fixed ratio of (rect. PE, EΣ) to (sq. EH). And this already constructs for us the main tool of the solution: the ellipse. The main idea had been found: all that remains is to force the ellipse and the central scheme of similar triangles to come together and intersect in some definite way. The way Diocles found was to introduce an extra hyperbola, and to rotate and transform slightly the ellipse, until they coincide at the point Ξ and, with this, we have the problem solved.

What kind of a solution is that? How does it compare to the solutions we have seen so far, by Archimedes and Dionysodorus? We can now quickly identify the objects used by Diocles, as "geometrical" or "quantitative" in character. The central scheme of similar triangles is geometrical, and similar in character to Archimedes' rectangle: it is a tool for presenting proportions concretely, as relations obtained for configurations of lines. The ellipse, on the other hand, is a *forced* conic section, not a natural geometrical object. Its only meaning in the proposition is that it happens to fulfill a certain metrical condition: it thus resembles the parabola in Dionysodorus' solution. The hyperbola, on the other hand, exactly like that of Archimedes', is closely related to the geometrically meaningful scheme of similar triangles, and serves here, as it did there, the configurational function of aligning points on a line. The final thought of the solution – the transformation of the ellipse – is based on a deeply quantitative understanding of the ellipse, as an object strictly defined by a quantitative value – the area of a rectangle that can be transformed at will. Pythagoras' theorem itself is reduced to its algebraic meaning – as a tool for deriving functional relations in, as it were, the second degree. For there are no squares involved here, merely a virtual rectangle whose value is seen to double because Pythagoras' theorem is seen here to ensure a doubling for any object defined by *two* lines. The rectangle itself is virtual, because (as is so often the case in the theory of conic sections) it does not represent two contiguous and orthogonal points: lines ΣE, EP and TO, OY are not orthogonal, but co-linear and this, in fact, is necessary for Pythagoras' theorem to take effect in this case.

In short, the verdict is mixed: why should it not be? We saw already one Greek mathematical proof displaying a deeply geometrical sense of the conic sections and of the overall treatment – that by Archimedes. We saw another, where the conic sections and the very approach to the problem were rather quasi-algebraic, the objects taken as mere stand-ins for quantitative relations – that by Dionysodorus. And so, quite naturally, we now find that the two approaches could mix in the very same proposition. This in itself is meaningful: there are no deep conceptual taboos involved (as authors such as Klein sometimes tend to suggest). The Greeks could think of objects in terms of their configuration, or in terms of their quantitative relations – and they could mix the two approaches. In all probability, they never even stopped to distinguish between the two.

Yet there is no question that the problem, as solved by Diocles, is still a problem rather than an equation. To start with – as with Dionysodorus – there is no equation to be solved here. The main feature of Diocles' approach is that he sets out to achieve not a single condition, but a system of three overlapping conditions. Needless to say, authors such as Heath, say, presented this as if Diocles was here solving a system of three equations![101] But this misses a crucial point: the extreme artificiality of the conditions in Diocles' problem. While stated, finally, in its general form, the problem is such that it has no meaning on its own. One is just never

[101] Heath (1921) II.47.

interested in getting three points fulfilling the three proportions stated by Diocles, unless one had a very special reason to do so. Archimedes' statement of the problem is very different in this respect: it is couched in such general and simple terms, that one can think of it as standing, so to speak, on its own feet. But Diocles' terms are so contrived, that they can be understood only as a thin disguise for a special configuration. In this way, Diocles' problem is indeed deeply embedded in the problem of cutting the sphere. Diocles' diagram (very crowded as it is) does not reproduce the sphere, as Dionysodorus' did. But the very problem is at a very little remove from the original one, of cutting the sphere. As it were, one can imagine a watermark underneath Diocles' diagram. It has a sphere with its diameter AB and two cones with their base on the point E and their vertices at the points Z, H, all representing Archimedes' statement of the problem of cutting a sphere. Turn Diocles' diagram against the light, and you see this watermark: unless you see it, the diagram is valueless.

Why does Diocles' write the way he does – producing a problem, rather than an equation? At this point, it would be best to proceed by considering, side by side, the three solutions of the problem extant from Hellenistic Antiquity: the problem in the world of geometrical problems.

1.6 The world of geometrical problems

Each of the three propositions is, I suggest, a problem and not an equation, geometrical rather than algebraic. But I have also suggested that none is purely geometrical. In Archimedes' treatment, we saw several dualities – several traces of a quantitative approach. Dionysodorus' understanding of the conic sections is essentially quantitative, and Diocles' approach is mixed. So this is an important preliminary result: none of the mathematicians covered here seems to operate under any constraint, preventing even the traces of a quasi-algebraic approach. This is very different, then, from what authors such as Klein would make us think. There does not seem to be a big conceptual divide, separating ancients from moderns, so that a certain type of mathematical understanding was inaccessible to the ancients. Greeks were perfectly capable of a quasi-algebraic treatment – but, in practice, they happened to minimize it. To account for that, then, we should understand the nature of their mathematical practice.

Let us recall the three solutions. Archimedes' solution states the problem in terms of a single proportion of lines and areas. It then solves the problem through a system of lines aligned along similar triangles, fixed by a parabola and a hyperbola. The similar triangles concretize proportions. The parabola creates a linear image of relations between areas. The hyperbola aligns points on a line. Everything is conceived, then, in terms of geometric relations. Thus the proposition is felt to be a geometrical problem, understood in terms of configurations of specific lines that have to be brought into a particular order.

Dionysodorus' solution states the problem in terms of the cutting of the sphere. It then solves the problem through a combination of hyperbola and parabola. The hyperbola yields a proportion between four lines (easily transformed into a proportion between four areas), which the parabola then transforms, through an equality between areas, into another proportion – the one required by the problem. The solution is conceived as an abstract manipulation of proportions, but the configuration is closely modeled on the original sphere, and the main resulting proportion is not highlighted as the key to the solution: rather, Dionysodorus goes through a long argumentation unpacking the solution in terms of cones and segments of sphere. Thus abstract proportions are seen as the background to the sphere, which is the foreground of the problem: once again, the proposition is felt to be a geometric problem, understood in terms of configurations of specific lines that have to be brought into a particular order.

Diocles' solution states the problem in terms of three separate points satisfying three separate proportions, all easily understood in terms of the original problem of cutting the sphere. From then on, the picture gets complicated: the problem is slightly generalized (the line representing the radius is replaced by any given line), the solution itself is complex and heterogeneous in its conception. It is based on a scheme of similar triangles, an ellipse, a hyperbola, and a transformation on the ellipse. The triangles and the hyperbola are reminiscent of Archimedes' geometrical conception, the ellipse and its transformation are reminiscent of Dionysodorus' more abstract conception. Once again, however, abstract manipulations of proportion are the background to the main issue. The main issue – the foreground of Diocles' solution – is the alignment of different geometrical objects into a single configuration. This is obtained sometimes by qualitative properties (similar triangles, a hyperbola), sometimes by quantitative properties (ellipse, Pythagoras' theorem), but the goal is always the bringing together of separate geometrical objects. The reason why this should be the character of this particular solution is obvious: Diocles had created for himself a piecemeal problem made of three separate tasks and conditions, so that the main goal of the solution has to be the synthesis of those different components. The basic understanding, then, is that the terms of a problem give rise to geometrical objects, and the solution is about aligning those objects in whatever means are available. Once again: a geometric problem, understood in terms of configurations of specific lines that have to be brought into a particular order.

The most important observation, then, considering now all three problems, is that each has its own special character. And this, alone, rules out the possibility of the problem becoming an "equation." This is because an equation demands a single conception of a problem, brought into some canonical terms. It appears that the dynamics of ancient mathematical writing go against the emergence of such single conceptions.

Only one of the authors – Archimedes – states the problem in simple and general terms that suggest the conception of the problem in wider terms. This is what modern readers are looking for: the conception of the problem as an example of "a cubic equation." This is not how Archimedes' first readers understood him. The later authors went on to restate the problem in terms of the sphere (as Dionysodorus) or in ad-hoc terms of aligning three separate tasks satisfying three separate proportions (as Diocles). Each of the later authors must have felt that, in his way, he had found a more elegant way of stating the problem. Dionysodorus had stated it in the more natural terms of the geometry of the sphere; Diocles had stated it in the simpler terms of proportions involving lines only (all of them, incidentally, aligned along the single line ΖΣ). Of course,

Archimedes, too, must have felt – with very good reason! – that he had found an elegant way of stating the problem. He had stated it as a single, general condition. But, considering Archimedes in light of his ancient reception, we can see that, in ancient times, what Archimedes had found was simply one possible route to elegance. Generality as such was not an overriding goal – as indeed we can see from the fact that neither Archimedes nor, indeed, Diocles, ever mention explicitly the generality of their stated problems.

What was the overriding goal, then? Apparently, just that – to find an original and elegant way of stating the problem (and then, of course, of solving it). Nothing surprising about that: Greek mathematicians aimed at originality and elegance. But how much follows from that! Because, originality being so high on the list of desiderata, it overrides the desideratum of generality. To aim at generality for its own sake is to look for canonical representations of problems, which is to forgo to a certain extent your originality. The Greek geometer did not do so. Thus, it is natural that each of the three solutions we saw was *sui generis*. The Greek authors do not aim to allow their solutions to fit some structure of classification within which their work can be recognized. On the contrary: they attempt to blur the outline of the problem, to hide their dependence upon different approaches. Archimedes' generalized statement is conjured out of nowhere, to surprise the reader (this is so that the reader would not see the general form as a mere technical tool, forced upon Archimedes to simplify the terms of the problem). Dionysodorus hides the dependence of his analysis on that of Archimedes – and precisely for this reason foregrounds the geometrical setting of his problem. And Diocles offers us an analysis – that is, the illusion of a solution yielded, naturally, by the terms of the problem itself – which involves a contrived ellipse, created and transformed for reasons the reader truly cannot fathom. In fact, it is forced by the special terms Diocles had forced upon himself, essentially, *just* to be special.

The essence of a *problem* – in the sense in which we now try to differentiate it from an *equation* – is that it is stated in the particular terms of a particular geometrical configuration. And it is clear that each of the three solutions has very different particular terms. It is important, after all, that they *look* different. The three diagrams (apparently, those produced by Eutocius) are each dramatically different from the others. Eutocius, if anything, would act as a unifying editor (for which, more in the next chapter). The originals, then, would have to be just as different from each other. With the exception of a single horizontal line AB – the shadow of Archimedes' original sphere – each diagram goes its separate ways: a structure dominated by a rectangle, in Archimedes; a structure dominated by intersecting conic sections, in Dionysodorus; a complex structure of similar triangles, and a slanted ellipse, in Diocles. Each solution has a special configuration leaping out of the page, and determining its special approach to the problem.

The very difference between the solutions explains why they are all problems rather than equations. As it were, all equations are similar to each other; each problem is a problem in its own way.

Notice that this observation has to be put positively, not negatively. It would be a mistake to characterize the Greeks by saying, negatively, that they do not aim at producing elements fitting a larger classification. Such a negative statement would merely help us to fix the Greeks, retrospectively, inside our own larger scheme, opposing them with later mathematicians. But synchronically, the negative statement cannot allow us to see why the Greeks, themselves, followed the route they did. We need to state the nature of Greek mathematics positively: not in terms of what the Greeks did not (but others would) aim at, but in terms of what they did. Historical epochs should be understood not in terms of what they were not yet, but in terms of what they already were.

And it is clear what the Greeks were – what they had aimed at. They had aimed at originality.

Perhaps the following analogy might help to bring out this aspect of the Greek mathematical practice. In a famous argument, Walter Benjamin had suggested in *The Work of Art in the Age of Mechanical Reproduction* (among the essays translated and collected in Benjamin [1968]) that, historically, works of art were singular objects surrounded by an "aura," as Benjamin termed it, derived from their very singularity. An unreproduced and unreproducible painting is a very special object in the world. It is a volume of cloth and paint, surrounded by the wood of its framing, touched by a unique master. It stands in its space, and the sense that there is nothing else that could represent it except the object marks out this space, as it were, as a singularity in the universe and endows it with a certain inapproachability. Benjamin suggested that, before the age of mechanical reproduction had punctured the uniqueness of the artifact, aura was essential to art's meaning. The aim was precisely to create this special, irreplaceable object. Now, I do not suggest any direct mapping between the history of art and the history of mathematics (in particular, the process of mechanical reproduction does not have a bearing on the dynamics of change in the history of mathematics I shall outline in the remainder of the book). But the concept of the "aura" is very useful in that it helps to bring out a feature of ancient mathematics that is very surprising for a modern reader. I suggest that the ancient author aimed at providing his work with an aura – with a sense of uniqueness that defies subsumption under any general heading. Of course, as with a painting, one could always make comparisons and produce catalogues. All three solutions above were perceived by the ancients themselves to be solutions of the same problem – just as three different paintings could all be seen, by their authors, to represent the same scene (e.g., an annunciation). But each unique annunciation painting would have its own aura, endowed by the special individual characteristics of its master. This is what the audience of the work of art had expected from the painter – what was foregrounded, for the audience; and this is what the audience of the mathematical proof had expected, in Antiquity, from the geometer. Even though the solution was that of a given problem, perhaps already solved by others, the goal was to produce a special approach, one carrying the author's signature. This is the essence of the features we have seen so far: the unique diagram characterizing each of the solutions; the separate statement of the very terms of the problem in each of the appearances; the intentional blurring of the background to the statement of the problem. Each of those features serves to make the solution not merely different from others, already offered, but also unique. In a sense, it is incommensurable with other approaches – and so, with its "aura," it is somehow inapproachable. The solution comes armed with mechanisms that rule out its reduction into some general form, in which it might be directly compared with other solutions.

Why would that be the case? To understand this, we need to see the context of the Greek mathematical practice. For which, once again, the foremost evidence are the treatises themselves. How do the treatises relate to each other? What sort of communication space do they presuppose? This space has several surprising features.

First, Greek mathematical treatises make very little *use* of each other. Archimedes, for instance, who does refer to earlier works – refers then, always, to works *by himself* (he would mention names such as Eudoxus and Democritus, but only as general historical comparisons, not for specific mathematical use of earlier treatises). When Archimedes relies upon previous results he simply uses them, sometimes quoting explicitly, but never mentioning the authors of those earlier results. This is true even for the result of "cone = 1/3 cylinder" (the one due to Eudoxus and Democritus) which Archimedes not only uses but in fact explicitly quotes, following *Sphere and Cylinder* 1.16 – where the quotation makes no reference to the source of this result, simply stating it as if was some unnamed piece of common knowledge! The one limited exception is in the *Sand-Reckoner*, where Archimedes explicitly refers to Aristarchus' heliocentric hypothesis. Typically, however, this is a hypothesis which Archimedes only takes on board for the sake of hypothetical study (it is not as if Archimedes suggests that this hypothesis is *true*), and, most importantly, the explicit reference is immediately followed by an explicit criticism (Archimedes corrects Aristarchus for assuming that the circle of the earth's trajectory is at the same ratio to the circle of the fixed stars as a center bears to a surface of a sphere; clearly this could have been at most a careless expression, but one Archimedes fixes upon, so that the quotation of Aristarchus becomes immediately a *criticism* of Aristarchus).[102] Similarly, Archimedes hardly ever refers in positive terms to any living person[103] and instead keeps heaping praise on the dead Conon. Partly, this must reflect Archimedes' noble dedication to his deceased friend, but the effect is also to belittle the value of any of Archimedes' contemporaries.

[102] Heiberg (1913) 218.7–31.

[103] The one exception is Eratosthenes, praised in the introduction to the *Method*: I have argued in Netz et al. (2001) that this praise may be ironic.

Is this haughtiness on the part of Archimedes? If anything, Archimedes showed a restrained, dignified attitude: at least, he did not badmouth the dead. This was Apollonius' specialty. The *Conics* contain harsh criticism of Euclid's own *Conics*, whose achievement Apollonius belittles: Euclid's work, merely a glimpse of Apollonius'! (Introduction to book I, especially Heiberg [1891] 4.10–16. Bear in mind that – if we accept the traditional chronology – we must assume that Archimedes, implicitly, found Euclid's *Conics* quite sufficient); Conon himself, we learn in the introduction to *Conics* IV, was wrong in his proofs (Heiberg [1893] 2.15–17; so admiration of Conon was not universal), for which he was rightly criticized – so Apollonius – by one Nicoteles; this being however the only thing Nicoteles was right about (Heiberg [1893] 2.18–22). All in all, according to Apollonius, no one ever dealt with conic questions properly – before Apollonius himself came upon the scene. Apollonius judged, and was judged himself: he is referred to posthumously, in similar terms, by Hypsicles, writing in the second century BC (introduction to *Elements* XIV). Or consider Eratosthenes: the addressee of Archimedes' *Method*, the Alexandrian librarian and therefore, in a real material sense, the custodian of the Greek past. Is he more deferential towards his predecessors? Not at all: in a well-known letter to king Ptolemy, where Eratosthenes sets out his device for the duplication of the cube, he offers a review of past achievements in the field – to criticize them and show their weaknesses relative to his own achievement (Heiberg [1915] 88.3 ff., most clearly 96.16–18). But wait! A generation or so later, Nicomedes, in his own work on the subject, will concentrate his efforts on demolishing *Eratosthenes'* achievement! (Heiberg [1915] 98.2–11). Well within this tradition, then, is Diocles' explicit criticism of Archimedes, in the quotation made in section 1.5 above – and, in this context, we cannot but see an implicit criticism of Archimedes in Dionysodorus' avoidance of the Archimedean analysis.

The way in which Greek mathematicians approach each other is through challenging each other. This "challenge" aspect is most obvious in another set of mathematical texts, for which we have far less evidence: *open problems*, or explicit challenges. References to such open problems are ubiquitous in Archimedes' writings:

he seems to have preferred to send out, first, open problems, and only then to send the proofs themselves (such open challenges are mentioned in the introductions to *Sphere and Cylinder* II, *Conoids and Spheroids*, *Spiral Lines*, and in the *Method* itself, in the part of the introduction we have not translated above). The famous *Cattle Problem* belongs to the same context. Thus we should not think of the correspondence of Archimedes as a series of *publications* – Archimedes communicating to the world his newest ideas. Instead, a published work, of the kind we have extant today, is merely a stage in an ongoing intellectual tournament.

The rules of the tournament are set up as one goes along, and it is not clear how Archimedes understood them. On at least one occasion, he did something that clashes with our natural sense of justice: he set out impossible challenges (as he explains in the introduction to *Spiral Lines*). Some of his challenges, he explains there, were intentionally wrong (i.e. impossible). Most probably, Archimedes is telling us the truth, here at least: he did not just slip and then save face by claiming that he made the mistake intentionally (this is shown by Archimedes' argument that the very mathematical context in which the false challenges were made, was sufficient to show their impossibility – as it were, a safeguard Archimedes had built into his false challenge against the charge that the falsehood represented his own mistake). The purpose of the exercise, according to Archimedes, was that "those who claim to solve everything, but produce themselves no proofs, shall be refuted by their claim to have solved the impossible."[104]

In short, then, one can make the following rule: the most natural way in Greek mathematics in which a previous work could be mentioned (in introductions) or envisaged (in open problems) was polemic.

Thus the space of communication is an arena for confrontation, rather than for solidarity. The relation envisaged between works is that of polemic. A Greek mathematical text is a challenge: it attacks past mathematicians, and fully expects to be attacked, itself.

The strategy we have seen so far – of the Greek mathematician trying to isolate his work from its context – is seen now as both

[104] Heiberg (1913) 2.24–4.1.

prudent and effective. It is prudent because it is a way of protecting the work, in advance, from being dragged into inter-textual polemics over which you do not have control. And it is effective because it makes your work shine, as if beyond polemic. When Greek mathematicians set out the ground for their text, by an explicit introduction or, implicitly, by the mathematical statement of the problem, what they aim to do is to wipe the slate clean: to make the new proposition appear, as far as possible, as a *sui generis* event – the first *genuine* solution of the problem at hand.

If so, we can explain, historically, why Greek mathematics produced problems, and not equations. Seen inside the context of polemical mathematical practice it is natural that Greek mathematical works should aim to possess an individual aura, in the sense developed above. Mathematical solutions possessing an aura would naturally have the characteristics we have seen in this chapter: solutions that involve configurations of specific lines that have to be brought into a particular order, everything possessing a mathematical meaning through an individual diagram, created especially for the particular solution. Such solutions strike us as "problems" in a real geometrical sense, rather than "equations."

I therefore suggest that the context of mathematical practice determined a certain approach to the writing of mathematical treatises, which in turn determined the nature of the mathematical contents themselves. To corroborate this model of historical explanation, and to begin to follow a trajectory of mathematical change, we should now move to the next stage in the life of Archimedes' problem: in the work of Archimedes' commentator, Eutocius.

2

FROM ARCHIMEDES TO EUTOCIUS

The texts we have read so far come not from works extant under the names of Archimedes, Dionysodorus, or Diocles. They were handed down in a single work, extant under the name of a relatively obscure scholar: Eutocius of Ascalon. In the sixth century AD, Eutocius wrote a series of mathematical commentaries, of which one, the commentary to Archimedes' *Second Book on the Sphere and Cylinder*, is especially rich in mathematical and historical detail. Having reached Proposition four, Eutocius noted the lacuna in Archimedes' reasoning. He has (so he tells us) uncovered Archimedes' original text, which he then incorporated into his commentary. Finally, he added into it the solutions by Dionysodorus and Diocles. This, then, is our main source for the ancient form of the problem (we also happen to have the same solution by Diocles, preserved in Arabic translation).

Was Eutocius' work a mere record of the past, or did it make some original contribution to the history of mathematics? In this chapter, I argue that, already in the work of Eutocius, we can find mathematics making the transition from problems to equations. This comes at seemingly trivial moments, of little consequence in terms of their original mathematical contribution. Eutocius, without noticing this, occasionally happens to speak of mathematical objects that are rather like our quantitative, abstract magnitudes, and not the spatial geometrical objects studied by Classical mathematicians. He stumbles across functions and equations, *without ever thinking about it*. In other words, such objects were not the product of a new conceptual scheme, but of a new practice.

Of course, no one claims that Eutocius was a major creative mathematician. But he did write on mathematics – returning to the themes suggested by Archimedes' problem, in a world very different from that of Archimedes himself. First and foremost, Eutocius lived in a world in which *commentaries to Archimedes were written*.

I shall argue that Eutocius' position as an author whose texts fundamentally depended on previous authors – what I call, then, a *deuteronomic* author – in itself signaled a transformation of the practice of mathematics, leading to subtle transitions – of important consequences. In the course of his own writing, Eutocius appropriated practices he had found in the text of Archimedes. His use of such practices, however, was different from Archimedes', because of his different position as an author. The different nature of texts determined the different nature of mathematics itself.

Still, Eutocius' own mathematics was very limited in its ambition, and so we shall not see in this chapter the deliberate introduction of new, rich mathematical ideas. For this, we shall have to wait for Arabic science, to which we turn in the next chapter. In particular, we shall not see in this chapter any new solutions to the problem, simply because Eutocius did not attempt any.

The best way to see Eutocius' originality is by a close comparison of Eutocius' text with what he had available to him from Classical Greek mathematicians. This is, after all, the way Eutocius' text was meant to be read: side by side with Classical mathematics. The main argument of this chapter has to do, therefore, not with the absolute originality of Eutocius, but his relative originality. That is, we deal not with what Eutocius achieved, but in how far he deviated from Archimedes. This chapter is about Archimedes as much as it is about Eutocius himself. We first discuss, in detail, certain Archimedean texts and practices. We then compare them to their appropriation by Eutocius. We concentrate on two remarkable innovations by Eutocius. The first comes close to the idea of a functional relation; the second comes close to the idea of algebraic magnitudes.

In sections 2.1–4, we discuss in detail how Eutocius seems to come across something close to the idea of a functional relation. In section 2.1 we translate the key text: Archimedes' (or perhaps, partly, Eutocius'?) study of the conditions of solubility of the problem of the proportion with areas and lines. In section 2.2 we try to unravel the two components of the text: which is by Archimedes, and which is by Eutocius? In section 2.3, we concentrate on the Archimedean part and show, once again, its geometrical character and, finally, in section 2.4, we discuss the original, more algebraic character of Eutocius' contribution.

Section 2.5 is dedicated to a single expression, "The area on the line." Used by Archimedes, this expression, in itself, suggests a quasi-algebraic understanding of geometrical objects. I argue that this expression serves a special function in Archimedes, setting apart a special text: Eutocius, however, "normalizes" the expression and in this way moves much closer than Archimedes did to a strictly algebraic conception of magnitudes.

Bringing together the two lines of argument – sections 2.1–4, and section 2.5–section 2.6 offer a brief summary on the transformation of mathematics in the world of deuteronomic texts.

2.1 The limits of solubility: Archimedes' text

Having quoted Archimedes' solution of the problem of the proportion of areas and lines, Eutocius then went on to reproduce Archimedes' discussion of the limits of solubility of the problem. This is an especially interesting text, both in the terms of Archimedes' originality, and of Eutocius'.

Why discuss the limits of solubility? Let us remind ourselves of the problem. It requires the proportion (fig. 9).

$$\mathrm{XZ:Z\Theta::(square\ on\ B\Delta):(square\ on\ \Delta X)}.$$

ZΘ and the square on BΔ are given. If we now use the strange terminology discussed in the preceding section, we can translate the proportion into an "equation":

$$\mathrm{(Square\ on\ \Delta X)\ on\ XZ = (square\ on\ B\Delta)\ on\ Z\Theta}$$

(where "on" should be understood to correspond, roughly, to "multiplied by": see section 2.5 below).

As both the square on BΔ and the line ZΘ are given, the right side of the equation is given. Therefore we are required to cut a line ΔZ at X, so that the square on one of the segments, on the other segment, equals a given magnitude. It is clear that we can deal with magnitudes as small as we wish. Make the segment that is squared smaller and smaller, and it is obvious that you will begin to get the equivalents of match-like prisms, as thin as you like, eventually as thin as nothing. Or make the other segment smaller and smaller, and you will get the equivalents of flat, slice-like prisms, eventually, again, as flat as nothing. This is an intuitive, very modern approach, looking for the behavior of expressions as they approach certain values (I shall return to the non-Greek nature of such an approach in section 2.3 below). At any rate, the issue of a hypothetical lower bound is not mentioned at all in the original.

However, there is an upper bound. The magnitude derived by taking the square on one of the segments on the other segment cannot be enlarged indefinitely. At some point in the line we reach a maximum, the largest magnitude associated with this line by this manipulation. This is therefore a limit on the solubility of the problem. If the parameters yield a greater magnitude than that, the problem is insoluble. That there is such a maximum, and where it is found, is proved in the following text (Heiberg [1915] 140.21–146.28):

And it will be proved like this that, BE being twice EA, the <square> on BE on EA is <the> greatest of all <magnitudes> similarly taken on BA.

For let there be, as in the analysis, again: (a) a given line, at right <angles> to AB, <namely> AΓ, (b) and, having joined ΓE, (c) let it be produced and let it meet at Z the <line> drawn through B parallel to AΓ, (d) and, through the <points> Γ, Z, let ΘZ, ΓH be drawn parallel to AB, (e) and let ΓA be produced to Θ, (f) and, parallel to it, let KEΛ be drawn through E, (g) and let it come to be: as EA to AΓ, so the <rectangle contained> by ΓHM[1] to the <square> on EB; (1) therefore the <square> on BE, on EA, is equal to the <rectangle contained> by ΓHM on AΓ, (2) through the <fact> that the bases of the two solids are reciprocal to the heights.[2] Now I say that the <rectangle contained> by ΓHM on AΓ is <the> greatest of all <magnitudes> similarly taken on BA.[3]

(h) For let a parabola be drawn through H, around the axis ZH, so that the <lines> drawn down <to the axis> are in square the <rectangle applied> along HM;[4] (3) so it will pass through K, as has been proved in the analysis,[5] (4) and,

[1] Mathematically, Step g serves to determine the point M.

[2] *Elements* XI.34.

[3] The point E is taken implicitly to satisfy the relation mentioned in the introduction to the proof: "EB is equal to twice EA."

[4] This is the Greek formulaic way of stating that the parabola satisfies the following condition. For every point taken on the parabola (say, in this diagram, T): "sq.(TX) = rect.(XH, HM)." (The point X is obtained by TX being, in this case, at right angles to the axis of the parabola and, in general, by its being parallel to the tangent of the parabola at the vertex of the diameter considered for the property.)

HM is known as the "parameter" of the parabola.

[5] Refers back to Heiberg (1915) 134.10–13. The line of thinking is roughly this. Let us try to show that rect.(ZH, HM) = sq.(ZK). We have rect.(ΓH, HM):sq.(EB)::EA:AΓ,

produced, it will meet ΘΓ (5) since it is parallel to the diameter of the section,[6] ((6) through the twenty-seventh theorem of the first book of Apollonius' *Conic Elements*).[7] (i) Let it <=the parabola> be produced and let it meet <the line ΓΘ produced> at N, (j) and let a hyperbola be drawn through B, around the asymptotes NΓH; (7) therefore it will pass through K, as was said in the analysis.[8] (k) So let it pass, as the <hyperbola> BK, (l) and, ZH being produced, (m) let HΞ be set equal to it <=to ZH>, (n) and let ΞK be joined, (o) and let it be produced to O; (8) therefore it is obvious, that it <=ΞO> will touch the parabola,[9] (9) through the converse of the thirty-fourth theorem of the first book of Apollonius' *Conic Elements*.[10] (10) Now since BE is double EA ((11) for so it is assumed[11]) (12) that is ZK <is twice> KΘ,[12] (13) and the triangle OΘK is similar to the triangle ΞZK,[13] (14) ΞK, too, is twice KO.[14] (15) And ΞK is double KΠ, as well, (16) through the <facts> that ΞZ, too, is double KH,[15] (17) and that ΠH is parallel to KZ;[16] (18) therefore OK is equal to KΠ. (19) Therefore OKΠ, being in contact with the hyperbola, and lying between the asymptotes, is bisected <at the point of contact with the hyperbola>; (20) therefore it touches the hyperbola[17] (21) through the converse of the third theorem of the second book of Apollonius' *Conic Elements*. (22) And it touched the parabola, too, at the same <point> K. (23) Therefore the parabola touches the hyperbola at K.[18] (p) So let the hyperbola,

and through similarity of triangles we can get EA:AΓ::ΓH:ZH (use, e.g., *Elements* VI.2). The combination of the last two proportions yields rect.(ΓH, HM):sq.(EB)::ΓH:ZH, or (through *Elements* VI.1) rect.(ΓH, HM):sq.(EB)::rect.(ΓH, HM):rect.(ZH, HM). This yields sq.(EB) = rect.(ZH, HM) (*Elements* V.9). Now notice that EB = ZK (through *Elements* I.34), so we can have what we have looked for: sq.(ZK) = rect.(ZH, HM). This certainly shows that the point K satisfies the condition of the parabola. *Conics* I.11 shows that all points lying on a parabola satisfy its condition, but we do not possess the converse, showing that all points satisfying the condition of the parabola lie on it (which is what we require here). This may well be a lacuna, not in our extant Greek corpus, but in Archimedes' reasoning here. I shall return to this in section 2.3 below.

[6] Steps c, e, h.

[7] The reference is to *Conics* I.26 in Heiberg's edition. Such references were certainly inserted by Eutocius. I shall return to this in the following section.

[8] Refers back to Heiberg (1915) 134.18–21. The argument is based on the following property (not the defining property) of hyperbolas: in a configuration such as of the diagram before us, they always satisfy relations such as "rect.(AB, BH) = rect.(ΘK, KΛ)," with the rectangles being contained by pairs of respectively parallel segments intercepted between the hyperbola and its asymptotes. This is (in Heiberg's edition) *Conics* II.12. That rect.(AB, BH) = rect.(ΘK, KΛ) is in this case a simple result of *Elements* I.43. Once again, the converse is taken for granted.

[9] "Touch" means "be a tangent of."

[10] *Conics* I.33 in Heiberg's edition. All we need then are Steps d, o.

[11] This is the implicit assumption of the entire discussion.

[12] Step d, *Elements* I.30, 34.

[13] Step c, *Elements* I.29, 32.

[14] *Elements* VI.4.

[15] Step m.

[16] Step d, *Elements* I.30. Finally, 15 derives from 16, 17 through *Elements* VI.2.

[17] In the sense of "being a tangent."

[18] As far as the extant corpus goes, this is a completely intuitive statement. Not only in the sense that we do not get a proof of the implicit assumption ("if two conic sections have the same tangent at a point, they touch at that point"), but also in a much more fundamental

produced, as towards P, be imagined as well,[19] (q) and let a chance point be taken on AB, <namely> Σ, (r) and let TΣY be drawn through Σ parallel to KΛ, (s) and let it meet the hyperbola at T, (t) and let ΦTX be drawn through T parallel to ΓH. (24) Now since (through the hyperbola and the asymptotes)[20] (25) the <area> ΦY is equal to the <area> ΓB; (26) taking the <area> ΓΣ away <as> common, (27) the <area> ΦΣ is then equal to the <area> ΣH,[21] (28) and through this, the line joined from Γ to X will pass through Σ.[22] (u) Let it pass, and let it be as ΓΣX. (29) And since the <square> on ΨX is equal to the <rectangle contained> by XHM[23] (30) through the parabola,[24] (31) the <square> on TX is smaller than the <rectangle contained> by XHM.[25] (v) So let the <rectangle contained> by XHΩ come to be equal to the <square> on TX.[26] (32) Now since it is: as ΣA to AΓ, so ΓH to HX,[27] (33) but as ΓH to HX (taking HΩ as a common height), so the <rectangle contained> by ΓHΩ to the <rectangle contained> by XHΩ,[28] (34) and <the rectangle contained by ΓHΩ> to the <square> on XT (which is equal to it <=to the rectangle contained by XHΩ[29]> (35) that is to the <square> on BΣ,[30] (36) therefore the <square> on BΣ, on ΣA, is equal to the <rectangle contained> by ΓHΩ on ΓA.[31] (37) But the <rectangle contained> by ΓHΩ, on

way, namely, we never have the concept of two conic sections being tangents even *defined*. I shall return to this in section 2.3 below.

[19] In Step k it has been drawn only as far as K. I shall discuss the verb "imagine" in subsection 2.2 below.

[20] Refers to *Conics* II.12, already invoked in setting-up the hyperbola. For the theorem to apply in the way required here, it is important that the asymptotes are at right angles to each other (as indeed provided by the setting-out of the theorem).

[21] Notice the cut-and-paste technique, and its concomitant labeling procedure through opposite vertices. I shall return to this in section 2.5 below.

[22] Converse of *Elements* I.43.

[23] The point Ψ is the intersection of the parabola with the line ΦX. Since this line had not yet come into existence when the parabola was drawn, this point could not be made explicit then, and it is left implicit now, to be understood on the basis of the diagram – this, the most complex of diagrams! A surprising amount of the work of specification of objects is left, in Greek mathematics, to the diagram, and not to the text (see Netz [1999] chapter 1).

[24] *Conics* I.11.

[25] Archimedes effectively assumes that, inside the "box" KZHΛ, the hyperbola is always "inside" the parabola. This is nowhere proved by Apollonius. Greeks could prove this, e.g., on the basis of *Conics* IV.26. I shall return to this in subsection 2.4.

[26] This step does not construct a rectangle (this remains a completely virtual object). Rather, it determines the point Ω.

[27] Steps c, d, *Elements* I.29, 30, 32, VI.4. [28] *Elements* VI.1. [29] Step v.

[30] Steps r, t, *Elements* I.34.

[31] *Elements* XI.34.

The structure of Steps 32–6 being somewhat involved, I summarise their mathematical gist:

(32) ΣA:AΓ::ΓH:HX, but

(33) ΓH:HX::rect.(ΓHΩ):rect.(XHΩ)

(34) rect.(XHΩ) = sq.(XT)

hence (from 33–4) the result (not stated separately):

(34′) ΓH:HX::rect.(ΓHΩ):sq.(XT)

ΓA, is smaller than the <rectangle contained> by ΓHM on ΓA;[32] (38) therefore the <square> on BΣ, on ΣA, is smaller than the <square> on BE on EA.

(39) So it will be proved similarly also in all the points taken between the <points> E, B.

But then let a point be taken between the <points> E, A, <namely> ς. I say that like this, too, the <square> on BE, on EA, is greater than the <square> on Bς, on ςA.

(w) For, the same being constructed, (x) let ϙςP be drawn through ς parallel to KΛ, (y) and let it meet the hyperbola at P; (40) for it meets it, (41) through its being parallel to the asymptote;[33] (z) and, having drawn A′PB′ through P, parallel to AB, let it meet HZ (being produced), at B′. (42) And since, again, through the hyperbola, (43) the <area> Γ′ϙ is equal to <the area> AH,[34] (44) the line joined from Γ to B′ will pass through ς.[35] (a′) Let it pass, and let it be as ΓςB′. (45) And since, again, through the parabola, (46) the <square> on A′B′ is equal to the <rectangle contained> by B′HM,[36] (47) therefore the <square> on PB′ is smaller than the <rectangle contained> by B′HM.[37] (b′) Let the <square> on PB′ come to be equal to the <rectangle contained> by B′HΩ.[38] (48) Now since it is: as ςA to AΓ, so ΓH to HB′,[39] (49) but as ΓH to HB′ (taking HΩ as a common height), so the <rectangle contained> by ΓHΩ to the <rectangle contained> by B′HΩ,[40] (50) that is to the <square> on PB′,[41] (51) that is to the <square> on Bς,[42] (52) therefore the <square> on Bς on ςA is equal to the <rectangle contained> by ΓHΩ on ΓA.[43] (53) But the <rectangle contained> by ΓHM is greater than the <rectangle contained> by ΓHΩ;[44] (54) therefore the <square> on BE on EA is greater than the <square> on Bς, on ςA, as well.

(35) sq.(XT) = sq.(BΣ)
hence the result (not stated separately):
(35′) ΓH:HX::rect.(ΓHΩ):sq.(BΣ)
and, with 32 back in the argument, the result (not stated separately):
(35″) ΣA:AΓ::rect.(ΓHΩ):sq.(BΣ)
whence finally:
(36) sq.(BΣ) on ΣA = rect.(ΓHΩ) on ΓA.

[32] Step v, *Elements* XI.32.
[33] *Conics* II.13. [34] *Conics* II.12. [35] Converse to *Elements* I.43. [36] *Conics* I.11.
[37] Steps 40–7 retrace the ground covered earlier at 24–31.
[38] This is a very strange moment: an already determined point (Ω, determined at Step v above) is now being re-determined. I shall return to this in the following section.
[39] *Elements* I.29, 32, VI.4. [40] *Elements* VI.1. [41] Step b′.
[42] Steps w, x, z, *Elements* I.30, 34. [43] *Elements* XI.34.
[44] Step b′, *Elements* VI.1.
The implicit result of:

(52) sq.(Bς) on ςA = rect.(ΓHΩ) on ΓA, and
(53) rect.(ΓHM) > rect.(ΓHΩ), is
(53′) sq.(B$) on $A < rect.(GHM) on GA.

This implicit Step 53′ (together with Step 1!) is the basis of the next, final step.

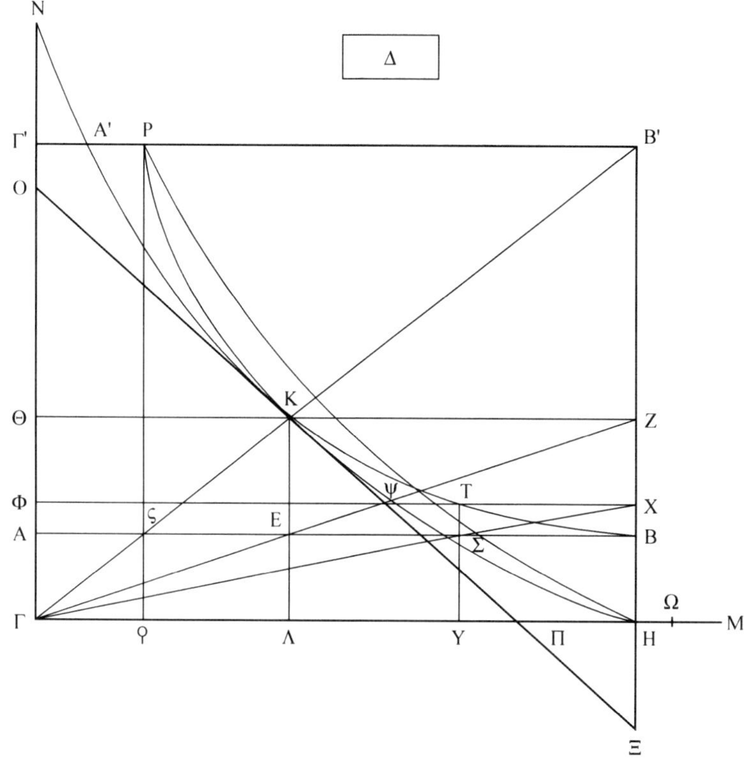

Figure 10

(55) So it shall be proved similarly in all the points taken between the <points> E, A, as well. (56) And it was also proved for all the <points> between the <points> E, B; (57) therefore, of all the <magnitudes> taken similarly on AB, the greatest is the <square> on BE on EA, when BE is twice EA.

2.2 The limits of solubility: distinguishing Archimedes from Eutocius

The entire passage on the problem of cutting a line was attributed by Eutocius to Archimedes on mere circumstantial evidence: the dialect is Doric, the terminology is early, the subject-matter is appropriate. There is nothing to add to Eutocius' argument. It is plausible, and one's preliminary assumption must be that whatever Eutocius had found was indeed by Archimedes.

This is not to say that what *we* read is exactly what Archimedes wrote. Eutocius complained bitterly about the poor shape of the manuscript he had come across, it was a very old manuscript, replete with mistakes (it seems that Eutocius means scribal mistakes).[45] Eutocius promises to transcribe this text "as it has been written,"[46] but with a few provisos. For one thing, he will correct the mistakes. For another, he will obliterate the very reasons which made him think this was by Archimedes, i.e. he will re-write the proof with modern terminology and in the dominant dialect.[47] But so far, this is not as bad as it may sound. The dialect difference is not such that it would have been noticed at all in the translation I have given above (the difference between written Greek dialects is mainly phonetic). The terminology involved is only that of the terms for conic sections. It appears that, before Apollonius, they were known as "sections of the right-angled cone," "sections of the obtuse angled cone," etc., while from Apollonius onwards they were known as "parabolas," "hyperbolas," etc.[48] Other than this, Greek mathematical terminology hardly changed through the centuries between Archimedes and Eutocius. So this again is no more than a trivial transformation. As for Eutocius' corrections of mistakes, these are potentially a more serious barrier separating us from Archimedes, but this should not be exaggerated. It must be realized that this kind of transformation occurs with every ancient text we now read. What is printed on the page is never what is written in the manuscripts. The editor will always consider some of the text of the various manuscripts to be bad copies of the original, and will correct them in the established "text," printed above the critical apparatus. It is a pity of course that Eutocius did not attach a critical apparatus detailing where he had differed from his source. But then he was no modern philologist.

These three transformations – critical corrections, dialect translation, and terminological standardization – are all innocuous. The trouble is that Eutocius clearly went beyond these. Consider for instance Step 6 above: "through the twenty-seventh theorem of the first book of Apollonius' *Conic Elements*." Apollonius was later

[45] Heiberg (1915) 132.1–11. [46] Ibid. 132.12. [47] Ibid. 11–15.

[48] The best treatment of the early history of the conics is Knorr (1982).

than Archimedes. Perhaps one may imagine that Archimedes made another reference there ("through the *k*th theorem of the *n*th book of Euclid's *Conic Elements*," say?), which Eutocius then collated with his Apollonius and transformed appropriately. But this would be sheer fantasy. We never have such references in texts which can be attributed to Archimedes with confidence.[49] Eutocius would not throw away such a precious reference to pre-Apollonian *Conics* (and would have mentioned such references as evidence for an early attribution of the text). So such steps as Step 6 stand for nothing in Eutocius' original.

The situation is as follows. Eutocius is in the business of writing a commentary to Archimedes. The format is textual. Eutocius follows Archimedes' text, and picks, here and there, some stretch of text for commentary. The commentary starts by quoting this stretch of text, and then goes on to make some comment on it. The text before us is all derived from a comment on a few lines by Archimedes in *Sphere and Cylinder* II.4 (where Archimedes had promised to add the lost appendix), namely Heiberg (1910) 190.22–192.6. Eutocius quotes this (Heiberg [1913] 130.2–17) and then his "comment" on this stretch of text gets him from 130.17 to the end of 176, of which 132.19 to the end of 146 are (possibly) the quotation of the manuscript found by Eutocius (140.21 to the end of 146 are the text translated in section 2.1 above).

Now bear in mind that the structure of text-comment is not allowed by Eutocius to be recursive. From 148 onwards, he goes on to make general observations (I shall quote some of them in section 2.4 below), and to make further quotations, from Dionysodorus and Diocles, with general comments following those quotations again. But he will not resume the structure of local quotation followed by local comment. He will not take some bits of the proof by Archimedes quoted by him, and quote them again for the purpose of a local explanation.[50] Avoiding such recursive commentary, Eutocius adds brief scholia to the quoted text, such as

49 We have one such reference, to Euclid's *Elements*, in Heiberg (1910) 12.3. Unique, and mathematically false, this is almost certainly a late interpolation to the text.

50 The principle of non-recursion does not bar one from further quotations of the first source in the course of one's comment. In the course of any comment, Eutocius may freely refer to any bit of the text of the *Sphere and Cylinder*. What he is not allowed to do is to take another piece, quoted in the course of some comment, as the basis for further comments,

Step 6. Unfortunately he uses a script with only one "font," as it were (basically what we know as capital Greek letters) – hence no setting apart of some text by, say, italics. It is also probable that he used no punctuation marks.[51] The net result is that what Eutocius must have perceived as innocent local interventions look to us like serious textual contamination.

How badly contaminated is the text from this point of view? It is of course difficult to say, but one rule of thumb is as follows. A natural way of inserting such scholia is as afterthoughts, as backwards-looking justifications, with Step n justifying Step n-1 (as, indeed, Step 6 is, justifying Step 5). The number of such backwards-looking justifications is not very large, nor is their size. Their distribution through the course of the proof is interesting, and I shall return to discuss it in the following subsection. My impression is that on the whole Eutocius added few scholia – and marked many of them clearly, with the references to Apollonius.

So far, then, we have seen enough reasons to be wary. Eutocius acted as a critical editor, as a translator, as a standardizer, as a scholiast, without letting us know where exactly he did all this. But there is much that can be still kept as "Archimedean." If we chop off the backwards-looking justifications, we shall have Archimedes' own words, more or less, to the extent that any translation may represent them. But this is on the assumption that Eutocius acted only in the above capacities. What if he acted also as a creative mathematician, in addition to all of the above? Here is the real worry.

I am not trying to imply that the entire text may be a fabrication. This is where criticism becomes mere scepticism whose only merit is that it cannot be refuted. But there is a real worry, namely where does Archimedes end and Eutocius begin? As I have mentioned already, following the extensive quotation 132.19–146.28 Eutocius goes on to add his own general comments. But Heiberg's page 148 does not start off with anything like "so far, then, the quotation." It

making it locally the first source. This is like our avoidance of footnotes-inside-footnotes, although inside any footnote we may make references to other pages, indeed to other footnotes.

[51] On the use in Greek mathematical texts of such tools of writing as "fonts" and punctuation, see Fowler (1999) sections 6.2–3.

simply moves on to another mathematical observation, which it is plausible to ascribe to Eutocius and not to Archimedes himself.[52] The worry is that we may have already slipped unawares from Archimedes to Eutocius somewhere prior to the end of 146. The text does not signal where the transition takes place, and the main reason I have so far taken the ending of 146 as the point of transition is that this was Heiberg's judgment.

In the following I shall argue that Heiberg was wrong in this judgment (which he himself represented as no more than a plausible guess).[53] I shall argue that a part of what we have read in section 2.1 above was in fact by Eutocius, and not by Archimedes, but before that I shall discuss briefly the possibility that the entire section 2.1 is by Eutocius.

It must be admitted that, while over sceptical, this is not impossible. We have two descriptions of the lost Archimedean appendix. Archimedes himself, in *Sphere and Cylinder* II.4, promises that, in that appendix, he shall offer analyses and syntheses for both problems, that of cutting a line in the general case and that of cutting a line in the case arising from the cutting of the sphere.[54] Therefore Archimedes promises the following sequence: (a) general analysis and synthesis, (b) special analysis and synthesis. Eutocius, about to give the contents of the manuscript he has found, says:

> First the problem shall be proved generally, so that what he says concerning the limits of solubility will be made clear; then, it shall be applied to the results of the analysis in the original problem.

As this comes following the quotation from *Sphere and Cylinder* II.4, where Archimedes says that in general, the problem has limits of solubility, it is clear that "what he says concerning the limits of solubility" refers backwards to this passage in the *Sphere and Cylinder*. Furthermore, the application to the special case as we have it in Eutocius (from page 148 onwards) seems, as already mentioned above, to be by Eutocius himself. By the same token, a doubt arises concerning the limits of solubility. Eutocius does not quite say that they are quoted from his manuscript. The only clear implication he makes is that the general problem is indeed

[52] I shall quote this observation in section 2.5 below.

[53] Heiberg (1915) 148, n.1.

[54] Heiberg (1910) 192.5–6.

taken from that manuscript. There seems to be some suggestion that the original contained a different sequence, possibly with more material that Eutocius did not quote. Take this in conjunction with Archimedes' own description of the lost appendix, and philological paranoia begins to resurface. Did Eutocius really come across the genuine lost appendix? How much of the original did he keep? How much did he add?

But let us regain our composure. Eutocius himself was competent as a commentator, he was on top of his Euclid and Apollonius, in short, he was capable of following any argument, however complex, and of adapting it creatively. But had he invented anything as original as section 2.1 he would have been wild with pride, and woe the bulls of Ascalon. I am not making the admittedly weak (though not meaningless) argument that "Eutocius was incapable of such proofs." Rather, the argument is that we would be certain to hear much more of this, had Eutocius been creative at such a scale.[55] On page 206, for instance, Eutocius comments on another analysis by Archimedes and then notes (lines 11–12): "Having said this, he himself did not bring in the synthesis. But we shall add it," and the synthesis follows. To derive a synthesis from an analysis is not a trivial operation, but it is much less than deriving limits of solubility from a solution. The *e silentio* is not conclusive. There are other places, where Eutocius probably added material of his own without comment (though never on this level of originality). This seems to have been the case with the synthesis of Diocles' solution. In fact, in the following section I shall argue this for part of section 2.1 above. But the *e silentio* remains a very serious argument.

On the other hand, if section 2.1 did occur, substantially, in the manuscript found by Eutocius, we are back to the question who could have written it and, again, the balance of probability is that it is by Archimedes. The span between him and Apollonius is too short, the quality is too high. And the fact that he does not

[55] Fabio Acerbi suggests (personal conversation) the following scenario: that Eutocius has discovered himself the entirety of section 2.1, and then, instead of making the claim for himself as a mathematician (*I have discovered this unknown property*) has decided to make a claim for himself as a commentator (*I have discovered this lost Archimedes manuscript*) . . . As Acerbi himself notes, this is a wild scenario but I have to concede that this is a possible one – reminding us of how much the ground we tread upon is speculative.

state explicitly that the limits on solubility will be discussed in the appendix shows nothing, after all. He may well have considered the proofs of such limits to form a necessary part of the solution of the general problem.

As we begin to read section 2.1, we still read, more or less, Archimedes. For how long? Or, to be more specific: Is the Proof for the Case of ς by Archimedes?

You may notice a line in the diagram, passing through H, T, and possibly through P, cutting the lines BΣ, ZK at anonymous points. The text we have read so far makes no reference to this line, but its main characteristics are clear. First, it passes through T, and perhaps through P. Second, it is "inside" the parabola HKN (the parabola with axis ZH and parameter HM). Furthermore, the text does include the following two constructions:

(v) . . . let the <rectangle contained> by XHΩ come to be equal to the <square> on TX.
(b′) Let the <square> on PB′ come to be equal to the <rectangle contained> by B′HΩ.

Steps v and b′ taken together indirectly define a parabola around the axis ZH, passing through T, P, lying "inside" the parabola HKN (since HΩ<HM). This then is the unnamed line. A first oddity of the source: the diagram includes a meaningful line which is not mentioned by the text as we have it.[56]

This leads immediately to another oddity, that of the point P. It is determined by the point ς. Now the point ς is logically different from the point Σ (although the two serve in the same roles, as points whose related solid magnitudes are smaller than the maximum at E). Σ is introduced by:

(q) And let a chance point be taken on AB, <namely> Σ.

Whereas ς is introduced by (intermediate setting-out, following Step 39):

Let a point be taken between the <points> E, A, <namely> ς.

[56] This parabola will be explicitly introduced by Eutocius following the text quoted so far, so its identity will become certain. But it is significant that this proof as we have it makes no reference to it.

The absence of "chance" from the definition of ς is meaningful. In fact it is anything but a "chance" point. The point ς is indirectly determined by the point Σ, in the following way. From the point Σ we derive a point T, from which we derive a parabola passing through H and T, with its parameter HΩ. This is the ghost parabola above, the parabola that dares not speak its name. Point P is then assumed to be the second intersection of this parabola with the hyperbola. This is assumed both by the diagram and by Step b′ quoted above:

(b′) Let the <square> on PB′ come to be equal to the <rectangle contained> by B′HΩ.

Point P, finally, determines point ς. So one visible logical sequence is ς, taken at the start of the second part of the proof, determining (in this context) P, determining Ω. Another, earlier chain of causation was Σ –> T –> Ω. So is Ω "variable," a mere notional end-point of a mere notional parameter, to be imagined freely moving on the line HM? Clearly not – the ghost parabola does not allow this. Moreover, the strange wording of Step b′ is decisive:

(b′) Let the <square> on PB′ come to be equal to the <rectangle contained> by B′HΩ.

"Let the square come to be!" (Although both points P and B′ are supposed to be settled by now!) In fact the author is in a remarkable position. He has on his hands two separate objects, the rectangle B′HΩ, and the square PB′. He needs to calibrate them, to make them equal, simply to save lines on a diagram which is too crowded anyway. He therefore assumes gratuitously that P falls on this very specific point, the intersection of the ghost-parabola and of the hyperbola. Against the visible chain of determination

ς –> P –> Ω,

there is another, invisible, stronger current:

Ω –> P –> ς.

In general, it often happens that the specification of objects in Greek mathematics is left for the diagram, so that the textual specification is a subset of the diagrammatic specification. This is the case in general and there is no logical problem about it, as long as the diagram is clear enough.[57] However, what is almost incredible is that the specification of the text and specification of the diagram *clash*. One may perhaps have cases where the text is ambiguous in itself, where the syntax of the text must be completed by reference to the diagram. But here the case is different. The text is crystal clear, and specifies the point P in one way. The diagram, no less clearly, specifies the same point in another way.

Do we really need the point P? We need it for the second case, of course, from Step 39 onwards. But do we need it at all for the first case? It is introduced already there. And this is how it is introduced:

(p) So let the hyperbola, produced, as towards P, be imagined as well.

Until Step p, we had the stretch of hyperbola BK. In Step 23, immediately prior to Step p, it was established that the parabola is tangent to the hyperbola at the point K, so now we are asked to "imagine" the hyperbola extended to beyond K, "as towards P."

Imagination is an established operation in Greek mathematics. I have discussed it in Netz (1999) chapter 1, and I have shown there that it often has a precise signification. It is used when the object to be "imagined" is not visible in the diagram, either because it is not an object a diagram can represent directly (a sphere, for instance), or because it simply is not drawn.[58] So we have discovered another clash between text and diagram, another, wider crack. The text of Step p seems to assume that the hyperbola of the diagram ends at K, and that there is a free-floating point P indicating the location of the continuation of the hyperbola (i.e. indicating a location in the box "above" the parabola HKN, "right" of the line ΓN, "left"

57 Netz (1999) chapter 1.

58 Another possible meaning of "imagination" is that of an addition by a commentator. A commentator may say "let us imagine x added to the diagram," having actually drawn x himself. We shall see this in section 2.4 below. But this cannot be the meaning here, for the following reason. The first part of the proof is (to the extent that it is independent of the preceding general solution) alphabetical, i.e. Θ is introduced before K, which is introduced before Λ, which is introduced before M, etc. Hence P must have been introduced originally at exactly the point where we see it introduced, just before the introduction of Σ. Whoever wrote the section of the proof from Σ onwards, must have written this Step p as well where P was introduced.

of the imaginary continuation of the line KΛ). Such a free-floating point is in itself a bold innovation, but it is what the text demands. The diagram is different, and actually extends the hyperbola to P. It is probable that whoever drew the hyperbola as far as P, was not the one who wrote the text of Step p. But notice that without drawing the hyperbola as far as P, the second part of the proof is impossible. Therefore the same probable argument seems to show that the second part of the proof could not have been written by the author of Step p.

More than this, Step p comes at the wrong moment, if we take the proof as a whole, the second part included. What does Step p do? It is not the basis of any construction or argument in the remainder of the first proof. We do not need the extension of the hyperbola before Step x, in the second part of the proof. The whole argument concerning the parabola and the hyperbola being tangent at K, as well as the extension of the hyperbola, are relevant only for the second part of the proof. So why are they brought in so much earlier, before they are actually required?

But then bear in mind that Step 23 (on the hyperbola being a tangent to the parabola at K), together with Step p, yield that the hyperbola, even above the point K, is contained by the parabola. The position of point P, inside the imaginary "box" right of ΓN and above the parabola HKN, is indeed all we need. For we can already see that the hyperbola is contained by the parabola *below* the point K. So what Steps 23 and p together do is to complete the argument that the parabola contains the hyperbola.

The remainder of the first part of the proof does not use anything about the point Σ except the following fact: the perpendicular TΣ cuts the hyperbola at a point T, so that XT<ΨT. Where the point Σ is on the line AB is immaterial to this argument (except of course that it is not allowed to coincide with the point E itself). Step q, in fact, is explicit:

(q) And let a chance point be taken on AB, <namely> Σ.

Σ is taken anywhere on the line AB, not necessarily on the stretch EB. The only thing required for the argument to hold is that the perpendicular TX shall cut the hyperbola at such a point where it is contained by the parabola. But this, we saw, holds generally. And therefore Step p comes at the right moment, just before the end of the first part of the proof, which is in fact already a complete proof. The way to make sense of the location of Step p is to understand it as a constituent in an argument in which the second part of the proof is redundant. But we already saw, through the verb "imagine," that the second part of the proof was probably written by a different hand to that of Step p. The probabilities begin to accumulate. It is time to replace the cumbersome expression, "the author of the second part of the proof," by the simpler name "Eutocius."

The story suggested is this. Archimedes gave a general proof (the first part of the text). Its generality was left implicit. Eutocius had to show explicitly that the proof was general. However, Eutocius was probably unable to give a good meta-mathematical account of the generality of the proof. Instead, he simply added an adaptation of the proof to the case that is not explicitly covered by the first part.

So this is the story; now the evidence for it. First of all, let us try to make sense of the silent parabola HKP and, in general, of the fact that P is determined by the parameter HΩ, although it should have been independent, determining its own separate parameter HΩ'. There is no point Ω, because the points ϛ and P are interesting for us, not as any points, but precisely as the points for the stretch AE whose related solid magnitude is equal to that of the points Σ, T. As we shall see in section 2.4 below, Eutocius goes on at page 148 to note the symmetry of the line AB around the point E. For each point at the stretch EB yielding a certain solid magnitude, another at the stretch AE yields exactly the same solid magnitude. For this, he needs precisely what we have here – a parabola such as HKP, and in general a line ϛP determined by the parameter HΩ. He refers to the pair T, P explicitly as such a symmetric pair. In other words the choice of ϛ, P is inexplicable if we wish to have a completely general proof for the stretch AE, but it is required by the continuation of the text, which is probably by Eutocius.

Why do we ascribe the text from page 148 onwards to Eutocius? There are many reasons, but the most important is that it starts off as a break from the preceding text, with a language that we would not expect in Archimedes' own proof (we shall see that below). There is no such obvious break between the two parts of the proof, but the connecting words immediately following Step 39 (soon after Eutocius takes over) are sufficiently foreign to Archimedes' discourse: "but then," *alla dē*, an expression which is common in Greek discursive prose but is very rare in the special discourse of Greek mathematical proofs. It is used only once in the Archimedean corpus (of about 100,000 words).[59] For Eutocius, the transition into a formulation that he considered non-Archimedean was sufficient to mark off the remaining text. Within such a rigidly determined practice as that of Greek mathematics, the slightest variations are sufficiently meaningful.[60]

Other kinds of text that Eutocius must have considered self-explanatory were the references to Apollonius, which I have described above as the scholia added by Eutocius. Eutocius did not need to mark them off as non-Archimedean; any historically competent reader could understand that. But now notice the following. There is nothing that looks like such a scholion in the second part of the proof. Almost all the backwards-looking justifications occur in the first part of the proof. So do all the references to Apollonius. And this is now obvious. Eutocius would not need to add any scholia to his own text. Rather, he would make it as clear as he could to start with. This is best seen in Steps 42–3, in the second part:

(42) And since, again, through the hyperbola, (43) the <area> Γ′Ϙ is equal to <the area> AH,

which should be compared to, say, Steps 29–30, in the first part:

(29) And since the <square> on ΨX is equal to the <rectangle contained> by XHM (30) through the parabola . . .

Step 30 is a backwards-looking justification inserted by Eutocius as an explication of Archimedes' claim 29. Step 42 is Eutocius' argument for his own following conclusion 43.

Another interpretation of the lack of scholiastic material in the second part may be that, after all, the second part adds nothing new. It simply adapts the argument of the first part. I am not sure this

[59] *Sphere and Cylinder* I.11, Heiberg (1910) 42.23.

[60] I have discussed such effects in general in chapter 2 of Netz (1999).

is any reason for Eutocius not to have given the Apollonian references, as commentators give references where they are relevant, not where they are necessary. But the "triviality" of the second part of the proof and its complete dependence upon the first part is of course the main argument for being suspicious. As explained already, the first part sets itself up as proving the general case. So the second part is not necessary as a matter of logic. And it is not mathematically interesting. So why should Archimedes bother to give it? On the other hand, adapting the same proof from one case to another is certainly within Eutocius' mathematical competence. More than this, it is sufficiently unglamorous, which explains why Eutocius makes so little noise about his adaptation. We therefore see that this second part of the proof is such that Archimedes would not wish to give, but Eutocius would, and could.

Finally, those strange symbols.

The diagram for this proposition introduces signs from beyond the alphabet. These signs are:

ϛ, ϙ, Γ′, A′, B′.

To repeat a point made in the preceding subsection, the script available to Eutocius (and to Archimedes) consisted of a single set of characters, roughly our capital Greek letters. Hence these extra signs are not "letters of a different type." They move on from the alphabetical sequence to another set of symbols, this time numerical.

Greeks used the alphabetical sequence for numbers: A was one, B was two, etc. Their numbering system was essentially decimal, and therefore they wanted, to begin with, twenty-seven symbols (for the twenty seven numbers, 1, . . . 9, 10, . . . 90, 100, . . . 900). The Classical alphabet however had only twenty-four letters. To fill in the holes, three extra symbols were taken over from archaic Greek: ϛ for six, ϙ for 90 and ϡ for 900. Moving upwards, the same letters were reused, with indices added: A′ for 1000, B′ for 2000, etc. At least two of the three extra numerals are used in this text (ϛ, ϙ). Two or three extra signs are required, and A′, B′, Γ′ are brought in (it is not completely clear whether we should read in our text ϡ or Γ′. Nothing hangs on it, and I have followed Heiberg's Γ′). Thus these extra signs are best understood as digits – it is exactly as if 1, 2, 3, 4, 5 would have been used. In the thirteenth century the Archimedean corpus was translated into Latin by Moerbecke, who perceived this and – being a follower of the latest trends – used the Arabic numerals 9, 8, 7 for Eutocius' A′, B′, Γ′. So these extra symbols are symbols for numbers, used, however, not *qua* numbers, but *qua* symbols. These are simply the symbols which happen to be available, now that the alphabet has been exhausted.

But while this use of numerals is in itself coherent, it is also a break from the established practice. The objects which are labeled by numerals are thereby strongly marked. *These are precisely the objects of the second part of the proof.* The second part of the proof declares its foreignness by its use of foreign symbolism. Once again, we see how deviations from an established practice are meaningful in themselves.

Now, might it be objected that one needed to introduce those symbols, simply because the alphabet was exhausted? Not quite: Archimedes was choosy in the first part.[61] He also used more than the absolute minimum of labels. Point Y is inert. It is introduced as part of the labeling of line TΣY in Step r, but nothing would be lost by calling this line simply TΣ, and the point Y is never mentioned again. Similarly, point Λ, introduced at Step f, does nothing for the argument. The free-floating point P is much more meaningful, but is still in a sense redundant. Generally, Greek mathematicians do not signal directions through free-floating points. Finally where did the letter Δ go? It is in the completely redundant area next to the main diagram, a leftover from the earlier diagram for the solution of the general problem of cutting a line. There, Δ served to signal the area which was the parameter for the problem, but for the proof before us, we no longer need to know the parameters of the problem, they are no longer relevant. Instead of Bach we now play Beethoven, but the harpsichord player remains on stage, with nothing left for him to do. This is no efficient management of resources.

What is striking is how neatly the first part of the proof takes up the alphabet, going exactly from A to Ω. This probably was

[61] Archimedes did not use the letter I. Indeed, this is often ignored in Greek mathematics, but not always, e.g., it is routinely used in Archimedes' *Spiral Lines*.

Archimedes' intention. Eutocius, squeezed out of the alphabet, picked up numerals.

Let us conclude by repeating the main result. Archimedes' own proof ended with the first part, at the end of Step 38. It included no cases, in fact no explicit generalization. It therefore had a very different mathematical character. Again, we may see here the transformation of mathematics through the presence of commentary. We are now in a position to consider this in detail. Having distinguished the two components of the text – from Archimedes, and from Eutocius – we may now proceed and discuss the character of each. Once again, we may, through this comparison, see the route leading from problems to equations.

2.3 The limits of solubility: the geometrical character of Archimedes' approach

What was the nature of Archimedes' own treatment of the limits of solubility? We will not be surprised now to read, for example, Heath's description of this text, in his *History of Greek Mathematics*:

> [The result for the limits of solubility] takes the form of investigating the maximum possible value of $x^2(a-x)$, and it is proved that this maximum value for a real solution is that corresponding to the value $x = 2/3a$. This is established by showing that, if $bc^2 = 4/27a^3$, the curves $[x^2 = c^2/ay, (a-x)y = ab]$ touch at the point for which $x = 2/3a$. If on the other hand $bc^2 < 4/27a^3$, it is proved that there are two real solutions.

This should not be dismissed without discussion. Indeed, in this case the algebraic reading of Archimedes is particularly tempting. A reason to think of the curves, discussed by Archimedes, in terms of equations is that this seems to suggest a way of obtaining the result. Even when the proof is followed and its validity seen, it remains unclear how it could be obtained. How did you know *in advance* where this maximum holds? That the sections must be drawn in this way? While we cannot perhaps see immediately the solution in terms of the equations, we can at least understand how they might have been used in principle: some algebraic manipulation, and surely the maximum could be obtained.

This, then, may have been Heath's view – though he remains silent on the possible approach which led to the discovery. Zeuthen, on whose work Heath's discussion is closely based, gives no further indications on how the solution was obtained (although he goes one step further in the algebraization, and offers an equation for the straight line ΟΞ!).[62]

To understand the meaning of the curves, then, we must look for the way in which they may have been discovered: manipulation of equations, or geometric intuition?[63]

We have before us a line AB (fig. 11), and we are looking for the most efficient way of cutting it, producing the largest possible solid magnitude of the square on one segment on the other segment. Notice incidentally that we are not looking for a *unique* maximal efficiency – as we begin, we are agnostic about this. What is clear is that efficiency cannot be extended indefinitely, so there is a "boundary." Whether this boundary is reached by one cut or by many is something we shall find out.

We may begin by thinking of two chance cuts on the line, C and D. What is the relation between the solids they generate? This is the ratio between the two solids:

sq.(BC) on AC:sq.(BD) on AD.

This is difficult to see, so let us concentrate on the squares, which form part of this ratio:

sq.(BC):sq.(BD),

and here it is obvious to the Greek mathematician that we may move to something simpler, since squares are always proportional to definite lines on a parabola. So let us have a parabola – any parabola – with its axis *EB* perpendicular to AB.[64] By *Conics* I.20

sq.(BC):sq.(BD)::FB:GB,

[62] Zeuthen (1886) 242.

[63] In the following I simplify by separating the theorem on the limits of solubility from the problem of cutting the line. This is a justified simplification, but bear in mind that Archimedes himself took the diagram arising from the problem of cutting the line, and therefore gave a slightly more complicated and less general proof – maintaining, in this way, the aura of the special problem he had solved himself.

[64] To make it easiest, have BC = FH, BD = GI.

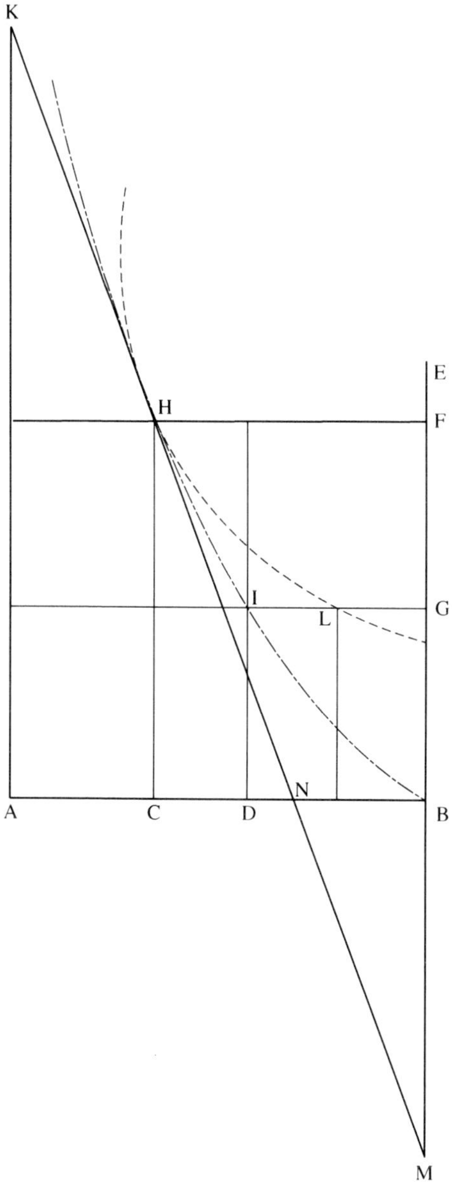

Figure 11

which means that the ratio of the two objects

sq.(BC) on AC:sq.(BD) on AD,

is the same as a much simpler ratio:

rect.(FB, AC):rect.(GB, AD),

or even better:

rect.(HCA):rect.(IDA).

Let us pause for a moment. We have now effected the crucial reduction of the problem, and its purpose is geometrical: to reduce a non-intuitive three-dimensional comparison into a comparison between two concrete areas. The third dimension is awkward, both for the diagram (as is obvious) and, derivatively, for the mathematical language (as we shall see in section 2.5 below). The role of the parabola is to allow a dimension-reduction. Instead of getting us away from geometry (into algebra), conic sections get us *into* geometry. But let us go on to follow the argument.

We have the two areas AH, AI in our hands. Can we compare them now? There is one obvious way to do this, based on another conic property, the equality of all rectangles on segments intercepted between the asymptotes and the hyperbola (*Conics* II.12). We have the asymptotes available, all we need to do is to choose a point. Choose H, and draw a hyperbola through it, with its asymptotes KA, AB. The only question now is where this hyperbola cuts the line GI. If it cuts it to the right of I, nearer G (as in the diagram, at L), then the point H yields a greater solid than the point I. And this is because the rectangle AL (equal to the rectangle AH through the property of the hyperbola) contains the rectangle AI – so simple, so geometrical. Similarly, if the hyperbola cuts the line GI to the left of I, further away from G, the generated solid will be greater at I; and if the hyperbola cuts GI at precisely the point I, the two points generate equal solids.[65]

We now have a firm grasp on the theorem, and we can state a condition. We are looking for a point, such that the hyperbola drawn

[65] This last simple argument is bypassed by Archimedes, who takes a somewhat more complex course. This is essentially because *his* parabola, a relic of the general problem, *cuts* the given line, instead of having its vertex on its end-point.

through it, with the asymptotes KA, AB, shall always be (left of the line EM) to the right of the parabola. It should be contained by the parabola in the relevant section of the diagram. So of course we are looking for a point where the generated hyperbola will be a tangent to the parabola.[66] In other words, the parabola and hyperbola should have the same line as tangent at this point.

Let us have this tangent at point H (which shall now be assumed to produce this problem). This tangent is then the line KM. Can we determine this line independently? Let us see, what do we know of tangents to conic sections? We reach for our tool-box,[67] and find there the following:

Conics I.33 tells us that, if KM is a tangent to the parabola

$$MB = BF \text{ (or } MN = NH),$$

and *Conics* II.3, that if KM is a tangent to the hyperbola

$$NH = HK,$$

and therefore, if KM is a tangent to both parabola and hyperbola

$$MN = NH = HK, \text{ or:}$$
$$BN = NC = CA,$$

and now we have shown that the greatest solid is generated at exactly one third the way. A simple way of discovery, corresponding in outline to the proof as we have it.

Which is a very satisfying way to discover this result. The point is one third the way not for some obscure quantitative manipulation, some black box of a calculation from which the number 1/3 emerges, but for the reason that we have here an equality between three segments. The remarkable thing is how little quantitative information about the conic sections is required. We do not ask for the precise parameter of the parabola, we know nothing about the parameter of the hyperbola. We have looked at a few interesting geometrical relations, involving various areas and lines: a geometrical play, acted upon the stage of conic sections.

[66] There is a crucial "topological" assumption here, to which I shall return below.

[67] By "tool-box" I mean (following Saito [1997]) the system of results with which the working mathematician is closely familiar (see also Netz [1999] 216–35).

Once again, then, we find that Heath's presentation was wrong, not only in that it made misleading implications for the mode of discovery, but for a deeper reason. Its conic sections were not the same as Archimedes', because he has lost the generality of the conic sections. Suppose one had given a proof of Pythagoras' theorem, where the sides and the squares are interpreted by algebraic equations involving Cartesian coordinates. This is possible, but it is a false and weak interpretation of the theorem, for it loses the general nature of the triangle of the purely geometrical proof. A triangle is not a quantitative object, its position and magnitude in the Cartesian space are not part of its essence. We still think of triangles in this way, because we have not forgotten the purely geometrical theorems for triangles, but we have learnt the Cartesian approach to conic sections, and have unlearnt the purely geometrical approach. We no longer share Archimedes' tool-box of geometric results for the conics, while we possess another, algebraic tool-box. And thus we find it difficult to see conic sections in the general, geometrical way Archimedes did.

In the proposition as offered by Archimedes, the sections are still adapted to the parameters which were earlier used in the solution of the general problem, but the main line of thought remains clear, and this is that sections of this general type will have a common tangent at one third the way, independently of their precise parameters. When Heath says that "this is established by showing that, if $bc^2 = 4/27a^3$, the curves [$x^2 = c^2/ay$, $(a - x)y = ab$] touch at the point for which $x = 2/3a$," he simply does not see the argument. Not because he is mathematically incompetent, but because his sight is obfuscated. He sees the wrong conic sections.

We see how Archimedes proves, and this is one aspect of a mathematical object, how do you prove with it? No less important is what you do not prove. There are a few characteristics of conic sections that are taken for granted here. For example, the uniqueness of a conic section with a given construction (i.e. the converse to *Conics* I.11, etc.). The proof of such results would be equivalent to another tacit assumption made here, that just because T is between Ψ, X, therefore ΨX>TX. Such spatial intuitions (known as *Pasch axioms*) were never proved in Greek mathematics. Further, there is a main assumption which is never stated, let alone proved. How do we know that, because they are tangent at one point, the hyperbola is always "inside" the parabola in the "box" NΓHZ? For after all conic sections may be tangents at one point, and then intersect at another. Archimedes (and Eutocius) simply take the assumption mentioned above as obvious. They share a certain intuition of conic sections as spatial objects; they know how they *behave*. It is such a spatial intuition that drives the argument.[68]

Once again, then, we see, following Klein and Unguru, a great divide separating ancient and modern mathematics. In the following section, we may see an example where – without breaking any new conceptual ground – a much more "algebraic" argument was given by Eutocius, following on his continuation of Archimedes' text.

2.4 The limits of solubility: Eutocius' transformation

Archimedes states a maximum. But does he approach it as a *maximum* – as a *limiting point*? The language of the calculus comes naturally, and indeed allows a very elementary approach to the problem. Starting from

$y = (a - x)x^2$ (inverting the segments for greater simplicity), or

$y = ax^2 - x^3$, we get the differential

$y' = 2ax - 3x^2 = x(2a - 3x)$,

[68] A Greek proof can be given. I sketch a possible argument, using the diagram of subsection 2.3.

1 Conic sections which are tangent to each other at one point do not meet each other at more than two more points (*Conics* IV.26).
2 Any "escape" of the hyperbola from the parabola inside this box must be compensated for, e.g., if at some point above K the hyperbola breaks free of the parabola, and approaches the asymptote GN, it must cross the parabola again so as not to touch the asymptote. So the hyperbola will then exhaust its two tickets within the box NGHZ.
3 This is impossible, however. For the hyperbola must cross the parabola at least once more, outside this box, right of the line HZ. The hyperbola will get nearer and nearer the asymptote GM whereas the parabola, symmetrical around the axis HZ, will retreat from the line GM. The two opposite movements must meet, so the hyperbola must keep one ticket free to be used there.
4 So it cannot exhaust its tickets inside the box. So the hyperbola cannot cross the parabola twice inside the box; hence it cannot cross it even once.

whose zero values are

$$x = 0, 2/3a,$$

hence the maximum at one third the way. Which of course only serves to show that Archimedes thought very differently. For one thing, we do not even need the conic sections now.

Still, the search for a maximum does look like a study of the properties of the curve. Is there anything of the kind in Archimedes? Does he have a sense, e.g., of how the maximum is *approached*? Does he see the line as the setting for a gradual motion, towards, and then away from the maximum point?

Here the conclusion of section 2.2 above becomes important. Once it is understood that the point Σ is taken as a completely general point, and that the first part of the proof was meant to be self-sufficient, the proof changes its nature. For now we no longer have a sense of the systematic relationship between the different points of the line. All we have is the relation between the point E, on the one hand, and all the rest (represented by the arbitrary Σ), on the other hand. The family of the different possible solid magnitudes generated by cutting the line was left uncharted. The proof is general, yet it does not yield any information about the continuous behavior of points along the line. But Archimedes was never interested in any such question: all he needed to show was that the problem of cutting a line so as to produce a solid magnitude, can be solved only up to a definite solid magnitude. The limits on the conditions of solubility are there to answer the yes-or-no question: is it soluble? They are not there as goals in themselves, showing the structure generated by various solutions.

So far, the proof as written by Archimedes. But it has already changed its nature by the addition of the second part of the proof. A certain structure was at least implied, that of a symmetry of the solutions around the point E. This, I have tried to argue, was Eutocius' supplement, leading on to his further comment. So let us read how his comment went on:[69]

[69] Heiberg (1915) 148.

Now[70] one must understand also the consequences of the diagram above. For since it has been proved that the <square> on BΣ, on ΣA, and the <square> on Bς, on ςA, are smaller than the <square> on BE, on EA: <therefore> it is possible to produce the task assigned by the original problem, by cutting the <line> AB at two points (when the given area on the given <line> is smaller than the <square> on BE on EA).

(a) And this comes to be, if we imagine a parabola drawn around the diameter XH, so that the lines drawn down <to the diameter> are in square the <rectangle applied> along HΩ;[71] (1) for such a parabola certainly passes through the <point> T.[72] (2) And since it <= the parabola> must meet ΓN ((3) being parallel to the diameter),[73] (4) it is clear that it cuts the hyperbola at some point above K, ((b) as, here,[74] at P),[75] (5) and <it is clear that> a perpendicular drawn from P on AB ((c) as, here, Pς), cuts AB at ς, so that the point ς produces the task assigned by the problem, (6) and so that the <square> on BΣ on ΣA is then equal to the <square> on Bς on ςA (7) as is self-evident from the preceding proofs.

So that – it being possible to take two points on BA, producing the required task – one may take whichever one wishes, either the <point> between the <points> E, B, or the <point> between the <points> E, A. For if <one takes> the <point> between the <points> E, B,[76] then, as has been said, one draws a parabola through the points H, T, which cuts the hyperbola at two points. <Of these two points,> the <point> closer to H, that is to the axis of the parabola, will procure[77] the <point> between the <points> E, B (as here T has procured

[70] Reading *dē* with the Greek manuscript, against Heiberg's *de* (which may have also been read by the Latin translator). This is not a trivial detail, since the manuscripts' *dē* is a more natural connector inside a stretch of discourse, whereas *de* is a more natural connector at the beginning of a new stretch of discourse. Hence my view, that Eutocius' own argument began earlier than this point, may help in keeping the manuscript's reading here.

[71] This is the ghost-parabola we saw above, finally given its own life: so its identification and its attribution to Eutocius become practically certain. Notice that it is "imagined," not in the sense that it is invisible in the diagram, but in the sense that it is a second-order, "imaginary" addition to the basic diagram (which is functional without it).

[72] From Step v of the proof, and then the converse to *Conics* I.11.

[73] Step 2 derives from Step 3, through *Conics* I.26.

[74] I.e. "in this diagram."

[75] Note that in this passage, the determination of P is unequivocally given by the point T. There is no longer any fiction that ς may determine P.

[76] The Greek runs, literally: "For if, on the one hand, the point between the E, B . . ." I have removed the meaningless "on the one hand" from the translation. It seems that as he started writing this, Eutocius thought he would go back to say the same about the segment EA (hence "on the one hand") but then realized this was redundant and, failing to proofread his text, it remained one limb short.

[77] The verb *heurisko*, better known to mathematicians for its first perfect singular used by an animate subject (*hēurēka*, translated "I have found," "I've got it"), commonly used in the infinitive with an animate logical subject understood (in the definition of goal inside problems: "*dei heurein* . . ." translated "it is required to find . . .", i.e. by the mathematician). Here, a third person present/future with an *inanimate* subject, the translation must be different, and mine is only one of many possible guesses.

Σ), while the point more distant <from the diameter will procure> the <point> between the <points> E, A (as here P procures ς).

There are two ways in which Eutocius is original. First, he describes a systematic relation holding in the line: the symmetry around the point E. Second, he has an explicit concept of a functional relation between mathematical objects. For him, that one point determines another is not an implicit feature, to be seen by an understanding of the proof, but an explicit relation, stated by him. But this is so original that even the term is new: *procure*. As mentioned in n. 77, my translation is no more than a guess – precisely because the term is original and is not a technical term of Greek mathematics. I suspect that Eutocius might have adapted a commercial sense of the verb, to express this functional relation,[78] hence my "procure." But of course it is impossible to interpret with any certainty such a singular text. The important thing is to notice the singularity.

Of course, Eutocius is still not completely modern. For instance, while he notes one structural property – namely the symmetry around E – he does not note another, namely the monotonic arrangement of the solutions. As we move towards E, the generated solids continuously become greater. This is obvious to us, imagining the solids as lying on the curve of a cubic equation, and seeing the properties of that curve. But Eutocius is very far from such a conception. He does not really move beyond Archimedes in terms of mathematical concepts. All he did was to adapt, mechanically, the proof. Superficially, it dealt with only one segment of the line and, starting from such a superficial observation, Eutocius adapted the proof to another segment. But because the adaptation was so mechanical, using the very same parameter of the parabola Ω, the adaptation made clear the existence of the symmetry around the point E. The sense of a functional relation between points reflects an awareness of this symmetry, no more. So mathematically Eutocius added very little – and yet he has moved so far away from Archimedes! One

[78] The idea is then that one point is determined by another, metaphorically, as a certain sum of money is determined by a certain article of merchandise.

almost feels that the very same comment may have been made by a modern commentator. Eutocius' conic section is an arena for equalities between points: it is thus, we may say, a sum of points, defined quantitatively. Thus it has become akin to the conic section of analytic geometry – which was never the case for the conic sections of Archimedes, Dionysodorus, or Diocles.

And the reason is simple: Eutocius is writing a commentary, and the very position of a commentary necessitates a transformation. To write about mathematical objects (as Archimedes did) is one thing. To write about mathematical arguments (as Eutocius did) is another. Limited as he was to the Archimedean mathematics, Eutocius must have given it a new meaning, without ever intending to. Archimedes says how one point on the line relates to another point on the line. But Eutocius also says how one point in the argument relates to another point in the argument (this is his remark on the symmetry of the line), and therefore the relation between points in the line becomes for him more like the relation between points in the argument (this is his concept of a functional relation between points).

Remarkably, we may see precisely the same transformation occurring, for the same problem, with Diocles' solution. As mentioned above, Diocles' solution is preserved also in an Arabic translation (as part of his *On Burning mirrors*). It is clear that the Arabic text may be closer, in some ways, to Diocles' text, than Eutocius' version is. Thus, Eutocius provides, in his text, several very elementary arguments that are omitted in the Arabic version, besides including the synthesis of the problem (which is a mere mechanical adaptation of the analysis he had from Diocles – and which is explicitly said, in the Arabic version, to be obvious once the analysis is known). In short, it appears that Eutocius had interfered in Diocles' text in a way directly comparable with his interference in Archimedes' text.

Diocles' solution, as a whole, does not give rise to overall considerations of functional relationships, the way Archimedes' text on the limits of the solubility does. But there is a small detail in the argument that gives rise to a similar transformation. This happens in Steps 16–17 of Diocles' proof (fig. 7):

(16) therefore the <rectangle contained> by ZEH is equal to the <rectangle contained> by PEΣ. (17) So through this, whenever P falls between the <points> A, Z, then Σ falls outside H, and vice versa.

As explained in my notes to this proof, the "vice versa" means that, conversely to what has been mentioned, also when Σ falls between B, H, then P falls outside Z.

Now consider the following. First, the Arabic text does not contain the words "and vice versa." Second, the point P happens to fall, in the manuscripts' diagram, between the points A, Z (with Σ falling outside H). In other words, Diocles' text merely argued for a property required for the diagram at hand: because P falls where it does, and because of the equality of the rectangles, the ellipse must be extended to beyond the point H. Diocles does not state a constant functional relationship, but justifies a particular diagram. Eutocius, then, goes on to add a small observation: a similar property would have held, even with a slightly different diagram. The diagram can be extended symmetrically: Σ is positioned to the right of H, because P is to the right of Z, but with Σ positioned to the left of H, P would then also be positioned to the left of Z. The observation, then, is at exactly the same – minimal – level of originality as Eutocius' extension of Archimedes' discussion of the limits of solubility. All Eutocius does is to add a case, symmetrically. But the addition of the case thereby changes the meaning of the argument: instead of a special observation on a special configuration, the text, transformed by Eutocius, sets out the constant relationships between possible configurations. In the case of Archimedes' limits of solubility, Eutocius finds that the same problem can be solved at two points, symmetrically arranged around a limiting point. In the case of Diocles' Step 17, Eutocius notices a constant functional relationship between two areas of a given diagram. Both observations betray very little mathematical ambition. Indeed, Eutocius hardly tries to make it apparent that these are his *own* observations. But both lead to the same result. Lacking any special ambition – or any new set of conceptual tools – we see Eutocius stumbling, as it were, across the idea of the function. In the next section, we shall see him get even nearer to the idea of the equation as such: treating magnitudes as if they were strict quantitative terms.

2.5 The multiplication of areas by lines

In Chapter 1 above we noticed various ways in which ancient texts – and, in particular, Archimedes' – display a certain duality: while geometrical in character, they also suggest the possibility of a more algebraic reading. One example of this was Archimedes' move to a more abstract statement of the problem, as a general proportion statement. Another was Archimedes' embedding of the solution within a lattice of orthogonal lines, suggesting the reading of the conic sections as the "curves" of analytic geometry, satisfying certain quantitative properties. In both cases, the suggestion was real enough, mathematically, but we could also see it as an incidental consequence of the real forces shaping Archimedes' text, always geometrical in character. The more abstract statement made sense, in that it made the geometrical problem more tractable, and the orthogonal lattice was useful for the geometrical preservation of ratios, and for the conservation of equalities between segments of lines.

I thus suggested, on the basis of such evidence, that the traces of a possible algebraic reading are just that – traces. The text is, simply, geometrical. The extreme complexity of the problem, however, compels Archimedes to deviate somewhat from the standard features of a Greek geometrical text. The duality of Archimedes' solution is comparable to that we saw in the solutions of Dionysodorus and Diocles, the first relying on a strictly quantitative understanding of the conic sections, the second combining a quantitative understanding with a more qualitative one, the more quantitative approach forced, in both cases, by the sheer complexity of the problem. The authors are simply brought to a point where, to move further, the solution has to use properties that are artificially introduced into the situation, without a clear geometrical basis. In short, we see that it is impossible to keep throughout a *natural* geometrical sense of the objects involved. In a complex problem, some artificial juggling is required, and, from a later perspective, this would appear as the traces of incipient algebra.

All of this, then, can be accounted for. But it is more difficult to make sense of another, final way in which Archimedes' text moves beyond a geometrical understanding of the objects. So far, we have seen how the complexity of the problem *compels*, to a certain extent, a more quantitative approach. But there is also a major way in which Archimedes' text, very surprisingly, makes a deliberate *choice* to deal with objects as if they were quantitative in nature. This choice, more than any other feature of Archimedes' text, points forwards towards a more algebraic understanding of the problem. Its later appropriation by Eutocius, in particular, would make Eutocius' text appear truly algebraic. We therefore postponed discussion of this feature of the text until now.

The feature is as follows. Archimedes' text contains many occurrences of expressions whose original wording is, e.g. "*to apo tēs AB epi tēn ΓΔ*" or, in more general terms:

(1) *to apo tēs* {two Greek letters} *epi tēn* {two Greek letters}
"The <square> on the <line> {two Greek letters} *epi* the <line> {two Greek letters}"

In other words this is an expression composed of three constituents:

1 The Greek mathematical formula for a square ("the on the {two letters}"),[79] followed by
2 The preposition *epi*, followed by
3 The Greek mathematical formula for a line ("the {two letters}").

In several cases, the first constituent (square) is replaced by the Greek mathematical formula for a rectangle ("the by the {two letters}{two letters}"). In the most general form, then, this is an expression composed of a formula for some two-dimensional figure, followed by *epi*, followed by a line:

(2) Figure *epi* line.
This is what I translate by

[79] Note that this formula (like most Greek mathematical formulae) is elliptic: the main noun, "square," is elided, and is understood (in the Greek) from the article.

(3) Figure on line.[80]

This composite expression is used as a description of a mathematical object, standing in standard mathematical relations. For instance we have seen expressions which may be rendered as:

(4) (figure$_1$ *epi* line$_1$) > (figure$_2$ *epi* line$_2$).

We are left with the question, what does the preposition *epi* denote?

First of all, notice that we should feel a sense of shock with this use of *epi*. At first glance this is a multiplication, pure and simple. An area is multiplied by a line, and the result is some mathematical object which can then be greater or smaller than others. These, then, are algebraic expressions. This is how Heiberg (1913)[81] took them in his Latin translation:[82]

(5) {square on line AB} *epi* {line ΓΔ}
becomes
(6) $AB^2 * \Gamma\Delta$.

Now we know – thanks to the work in the tradition of Unguru (1975) – that Heiberg's expression "AB^2" is misleading, and this is because the original Greek made a reference to a concrete square, not to the abstract operation of *squaring a quantity*. But can we say the same of Heiberg's expression "*?" What does it stand for, in concrete geometrical terms? Mugler (1972), pre-Ungurian but sensing the oddity of the expression, offers at one point the translation:[83]

(7) *La figure solide, produit du carré sur AB par ΓΔ.*
(The solid figure, produced by the square AB on ΓΔ.)
But in general he offers a rhetorical equivalent of Heiberg's translation:
(8) *Le produit du carré sur AB par ΓΔ.*
(The product of the square AB on ΓΔ.)

[80] The English preposition "on" is chosen to represent the Greek "*epi*," for its equal potential ambiguity, and no theory lies behind it.
[81] 141 ff.
[82] Followed by Heath (1897) 69 ff. and by the Loeb translation, Thomas (1941) II.149 ff.
[83] 94 (translating 94.3–4).

Phrase (8), unlike (7), is a translation of the Greek. This is because the original Greek makes no reference to any "solid figure." One thing should be clear: this is not a description of a prism, for no such prism is mentioned in the text. True, Archimedes uses this expression as if it denoted a geometrical object,[84] but this is not surprising in a geometrical context, and merely underlines the fact that no reference to a geometrical object is explicitly made.

Greek could easily accommodate elliptic expressions such as, let us say:

(9) The by the on the AB and by the ΓΔ
standing for
(10) The <prism produced> by the <square> on the AB and by the <line> ΓΔ.

Greek mathematical language is highly elliptic. Often, nouns are dropped, with only their articles left,[85] but this very tendency to elide makes every word meaningful. Wishing to refer to a solid figure, Greeks would name it. They would probably drop everything except the article, but the article would be there. But here there is nothing referring to a solid: just "area *epi* line."

Moreover, *epi* does not refer to the construction of a solid from an area and a line. In the expression suggested above, I have used "produced by" to express this idea of construction. But *epi* does not have this meaning. In fact we have a clear sense of what it might mean, and this is because it is often used in calculations, in expressions of the form:

(11) number *epi* number
elliptic for
(12) number <multiplied> *epi* number.[86]

[84] See especially Step 2 of the proof, in section 2.1 above, where the expression is taken as equivalent to "solids" (this Step 2, however, may well be by Eutocius, and not by Archimedes).

[85] This ellipsis is possible, because Greek has a rich morphology of the article, so what I translate monotonically by the English form "the" is expressed, in the original, by several different forms which are easy to differentiate.

[86] E.g., in Eutocius' commentary on Archimedes' *Sphere and Cylinder* Book II, Heiberg (1915) 122.7–8.

Moreover, Greek has a case-system: a preposition such as *epi* may be followed by a noun in the genitive, the dative, or the accusative, and the meaning of the preposition will be different according to the case taken by the noun. Most often in Greek mathematics, *epi* is used in various shades of meaning of "on," taking the genitive: e.g. "a point taken *on* a line," "a line added to another *on* a line with it" (i.e. the angle of the two lines is 180 degrees). With the accusative, *epi* is sometimes used in a spatial sense of direction. In this case, "on" takes the meaning of "towards" (or "as far as"): e.g. "a line being produced *towards* a point/line/plane." This may even result in a three-dimensional structure reminiscent of the object at hand, e.g. (*Elements* XI.35):

(13) "Let perpendiculars, <namely> HΛ, MN, be drawn from the points H, M on [*epi*: also translatable in context as "towards"] the planes <passing> through the <angles> BAΓ, EΔZ."

Here the sense of motion towards a spatial goal gives rise to a line built on a plane – the object we are looking at. But the preposition *epi* does not have the function required. It does not serve as a static description of the three-dimensional object resulting from the plane and the line; rather, it serves dynamically, to lead the act of drawing. If we were to take this geometrical *epi* as our guide, then, we would have the expression meaning something like:

(14) the area towards the line.

But this clearly is irrelevant here. So we have only one usage of *epi* with the accusative that may be relevant here, and this is, indeed, the arithmetical usage, *<multiply> by*. The linguistic evidence is therefore in favour of the translation

(15) figure <multiplied> by line.

In other words, Heiberg's "*" is vindicated and, with it, the sense that

(4) (figure$_1$ *epi* line$_1$) > (figure$_2$ *epi* line$_2$)

is an algebraic expression. So what shall we say? That geometrical algebra has been vindicated?

There are several reasons to feel uneasy about this. There is of course the fact that this is an *anomaly*. There is no question that in general, Greek geometrical texts are spatial, and not abstract. The criticism of "geometrical algebra" is a genuine achievement of the historiography of Greek mathematics. We should not forget its lessons. So such seeming counter-examples must be treated with caution.

More troubling still, it is difficult to give a coherent mathematical account of such a multiplication. There is no doubt now that the Greeks never had any concept equivalent to our concept of "real numbers." Numbers were positive integers. In some contexts of calculation, "fractions" (in a limited sense) were treated analogously to numbers. For instance, we can have expressions of the form

(16) fraction <multiplied> *epi* fraction.

But even there, the Greeks did not typically think in terms of our *rational* fraction, i.e. they did not think of every ordered pair of integers as constituting a fraction. What they had were mostly "parts," e.g. half, third, quarter, and their combinations.

The main concept that Greek mathematicians used instead of "fractions" or "real numbers" was the ratio. This was always the relation between two objects, and not a single object. Thus, it was very unnatural to say, e.g.:

*(17) The length of this line is the square root of two.
Instead, one would say:
(18) This line is to that line as the side of this square to the side of half this square.

It appears that for authors in the main tradition of Greek geometry there was no mathematical sense in ascribing to lines and figures the quantitative measures "length," "area," or "volume." In general, Greek mathematicians do not use expressions such as:

*(19) the length/area/volume of this line/figure is {number}.

This is because while such expressions could conceivably mean something in the special case where the number was what we call an integer, such expressions would break down, in the Greek context, with irrationals and, indeed, with most rational numbers as well.[87]

However, if there is no quantitative measure of this sort, what is it that Archimedes multiplies when he uses this expression? Or, to put it differently, did Archimedes not consider the possibility that the lines he discussed might (as we would express it) be irrational? From a mathematical point of view, the possibility must be open, because the proof is general, and is meant to apply over a continuum.

On the other hand, there can be little doubt that the expression is genuine Archimedes. As explained in section 2.2 above, the source used directly by Eutocius was probably very close indeed to an original Archimedean text, and while Eutocius has certainly transformed this source in several ways, he would have had no motivation for introducing the *epi* locution (a suitably elliptic prism-based locution would be easily understood by Eutocius, and would not have been more cumbersome than the *epi* locution). I add that we have the *epi* locution used repeatedly in Archimedes' *Sphere and Cylinder* II, the alternative proof to Proposition 8, (the penultimate proposition of the book).[88] I personally believe that this alternative proof is by Archimedes himself (it is radically original in many ways, which may explain why Archimedes would have been interested in offering such an alternative proof in addition to a more "standard" proof – while it is difficult to see who else was capable of and interested in producing such a proof, only to leave it as a gloss in the text of Archimedes!). At any rate, the expression was certainly used in this geometrical context, if not by Archimedes himself, then by some other highly competent Greek mathematician.

In what follows, I shall not deny that this expression is in a sense more "algebraic" than others we see in Greek mathematics (always, of course, taking "algebra" in the most elementary sense). What I shall do is to qualify this by discussing the following three questions:

[87] For the discussion above, see Fowler (1999) chapter 7.
[88] Heiberg (1910) 218.1 ff.

*Is the expression completely algebraic?
*Why does Archimedes use it?
*In what contexts is it used?

Is the expression completely algebraic?

I have said above that Heiberg's translation of *epi* by "*" is vindicated by the fact that the *epi* does not refer to the construction of any geometrical figure. However, of course, there is much more than this to the sign "*." In several ways, the use of *epi* in the expression before us is different from the algebraic usage.[89]

One of the features of "*" is its commutativity. In everyday multiplication, we assume a*b = b*a. At least, when "a*b" is a meaningful expression, so must "b*a" be. This is part of the significance of the expression "*." This is also true of the standard Greek way of using *epi*, when multiplying numbers. There, it is always possible to change the order of the numbers multiplied, and to derive a meaningful expression which is mathematically, though not syntactically, equivalent to the first one. Here, however, we have (in this text, and in other related sources in Archimedes and in Eutocius) at least around fifty tokens, all of the form:

(2) Figure *epi* line
and none of the form:
*(20) Line *epi* figure.

The last hypothetical expression is apparently undesirable. While there is some sense in which a figure can be *epi* line, this does not naturally extend to a line being *epi* a figure.

In fact, this *epi* is never used in any other context besides "figure *epi* line." Although the texts we look at often mention, for instance, rectangles, we never have expressions of the form

*(21) Line *epi* line.

[89] What I shall do in the following few paragraphs is to analyze the meaning of a mathematical sign not by trying to elicit the possible intentions attached to it by practicing mathematicians, but by looking at the textual configurations in which it may appear. In other words, I shall now analyze meaning not through intension, but through extension. In this I follow the method offered by Herreman (2000).

So this *epi* has an extremely narrow extension. It is limited to a single geometrical context. In a sense, this is the opposite of "*." The very reason we consider this last sign as "formal" or "abstract," is its open-endedness, its universal extension. "*" is blind to the contents of the multiplicands, and to this extent it is about form, not about content. But this geometrical *epi* is extremely content-sensitive.

Furthermore, even the arithmetical *epi* may be different from the sign "*" as we use it. I believe the following observation may hold for the arithmetical case as well, but my evidence derives from the geometrical case. This is the following: I have asserted above that the *epi* constructs a new mathematical object, which may then be inserted into standard mathematical relations, e.g.

(4) (figure$_1$ *epi* line$_1$) > (figure$_2$ *epi* line$_2$).

However, are we really justified in parsing the expression in such a way, i.e. in taking the relation ">" to hold directly between the two composite objects of the form "figure *epi* line?" Perhaps the correct parsing is different, and the relation ">" holds (albeit indirectly) between the two figures? So perhaps a better punctuation is:

(22) figure$_1$ (*epi* line$_1$) > figure$_2$ (*epi* line$_2$).

In this case we cannot decide between the two readings, but fortunately we have other mathematical relations, where the distinction is more obvious. My evidence now comes from Archimedes' *Sphere and Cylinder* Book II, alternative proof to Proposition 8.[90] This is a sentence where two expressions are said to be equivalent (two ratios are said to be the same). I print the copula "is" connecting the two ratios as bold, and I separate the two ratios by "|."

(23) *ho . . . tou apo AΘ pros to hupo tōn BΘΓ |ho tou apo AΘ* **estin** *epi tēn ΘH pros to hupo tōn BΘΓ epi ten ΘH.*

Word by word:

The <ratio>, of-the <square> on AΘ, to the <rectangle contained> by the <lines> BΘΓ |the <ratio> of-the

[90] Heiberg (1910) 220.3–4.

<square> on AΘ **is** *epi* the <line> ΘH, to the <rectangle contained> by the <lines> BΘΓ *epi* the <line> ΘH.

Meaning:

"The ratio, of the square on AΘ, to the rectangle contained by BΘΓ | **is** <the same as> the ratio of the square on AΘ *epi* ΘH, to the rectangle contained by BΘΓ *epi* ΘH."

A partially algebraic paraphrase of the mathematical content:

sq.(AΘ): rect.(BΘΓ)::sq.(AΘ) *epi* ΘH:rect.(BΘΓ) *epi* Θ*H*.

As can be seen in the word-by-word version, Greek has a different word-order from English. The copula, "is," occurs not immediately between the two ratios which it connects, but inside the second ratio. The most natural position for this copula in Greek is immediately following the first object of the second part. This is in fact where it stands. But what is this *first object*? The location of the copula is crucial. Had the "figure *epi* line" expression constituted a single composite unit, the copula "is" would naturally appear following that unit, and certainly could not break it into two the way it does here. In fact, the word "is" occurs *inside* the "figure *epi* line" expression: it comes immediately following the "figure" part. In other words, the syntax is that suggested above, as example 20. The "figure *epi* line" phrase is not a single object, but is a composite clause, with a noun – the figure – modified by the adverbial expression "*epi* line." So when we have an expression such as

(24) A *epi* B,

we do not thereby set up a third object, standing on its own, apart from the two objects A and B. It is not surprising therefore that we never have expressions such as

*(25) A *epi* B *epi* C.

I have stated above that, in the geometrical case, *epi* is not commutative. In both the geometrical and the arithmetical cases, the question cannot even be raised, whether *epi* is associative or not. That is, we cannot have the question whether

*(26) (A *epi* B) *epi* C = A *epi* (B *epi* C).

There is no such thing as "(A *epi* B) *epi* C" (no uninterrupted sequence of more than a single *epi*), and this is because, effectively, there is no such thing as (A *epi* B). This is because the *epi* phrase does not constitute a separate object. So here again we see a crucial distinction between *epi* and "*."

In this geometrical context, we see an operation which is neither commutative nor associative, and which only applies to the pairs <two-dimensional figure, line>. More precisely, this is not an operation at all. An operation takes two objects of a given domain to produce a third. But here we take the first object from one domain, the second from another, and we do not get a third at all. The grammar is completely unlike that of multiplication or of any other operation. So finally, *epi* simply is not "*."

But the difficulty remains, since this *epi* has no geometrical reference either. It projects no geometrical figure, and we must still account for this absence.

Why does Archimedes avoid a geometrical reference?

Generally speaking, Greek geometrical objects are compared and manipulated by two separate techniques, related to the duality we have referred to already several times above. Sometimes objects are considered as strictly spatial, and then they are put together and manipulated in space. Sometimes they are considered as satisfying certain quantitative relations. The first approach calls for the cut-and-paste technique, with expressions such as, e.g. (fig. 12):

(27) AE is equal to EH. Let BF be added in common. Therefore AF is equal to BH.

The second calls for proportion theory, with expressions such as, e.g.:

(28) As the line AB to the line CD, so the line CD to the line EF. Therefore the rectangle contained by the lines AB, EF is equal to the square on the line CD.

Figure 12

The two types of operations require different labeling procedures. In the cut-and-paste technique, it is only necessary that we refer unambiguously to the diagrammatic object. By signaling the borders of the object (e.g., as in the example above, through the two opposite vertices of a quadrangle) it is clearly marked, and can be diagrammatically manipulated. For proportion theory, on the other hand, another labeling procedure is required. Here the diagrammatic reference is much less important (in fact my example above did not require any diagram). What is imperative is that the labeling would refer to what may be called the "measuring constituents" of the object. For instance, the rectangle should be labeled through its two contiguous sides, the square should be labeled through its generating side, etc. It is these measuring constituents that are manipulated by the operations of proportion theory.

In the case of the most common objects of plane geometry, this bifurcation of techniques yields a bifurcation of acceptable labels. Rectangles may be labeled by either two opposite vertices (sometimes, all four vertices) or by two contiguous sides. But in general objects tend to belong to one or the other of these families. Greek mathematics employs a very narrow system of formulaic labels for objects. Each object has its preferred procedure of labeling. It is this rigid formulaicity of reference which makes the radical ellipsis of Greek mathematical expressions possible.

In the extant Greek mathematical corpus, three-dimensional objects divide in an interesting way according to these two labeling procedures. Cubes, like squares, are in the "proportion" family, named by their generating side. Cones and cylinders belong to the same family, and are often referred to as

(29) The cone/cylinder having the circle A as base, and the line B as height.

The reason for this is clear: cones and cylinders can hardly be manipulated by cut-and-paste techniques. Furthermore, this labeling by measuring constituents is unambiguous. However, prisms (and, in general, parallelepipeds) tend to be labeled, in the extent corpus, by their boundaries: sometimes by opposite vertices,[91] sometimes by opposite lines,[92] sometimes by opposite planes.[93] Part of the reason for this may be that such objects (being rectilinear) admit of cut-and-paste manipulations. More important, these are very complex diagrammatic objects, made of an intricate network of lines on the two-dimensional surface of the page. This complexity would have been aggravated by the schematic nature of Greek mathematical diagrams.[94] Such objects had to be named through their spatial boundaries, not through their measuring constituents – simply in order to be seen.

Whatever our explanation of the practice, the practice itself is clear. While in terms of the Greek language, the following expression is possible, and certainly would have been understood by Greek mathematicians:[95]

[91] E.g., *Elements* XI.39 (Heiberg [1885] 134.14).

[92] E.g., *Elements* XI.24 (Heiberg [1885] 74.6).

[93] E.g., *Elements* XII.4 lemma (Heiberg [1885] 162.18–20).

[94] See Netz (1999) chapter 1.

[95] I have mentioned this possibility above, in a slightly anachronistic form, as example 9.

*(30) The prism having the area A as base, and the line B as height, such expressions are rare in the extant corpus.[96]

In the Archimedean passages where our *epi* is used, Archimedes needs to discuss, effectively, prisms, and he needs to discuss them in the context of proportion manipulations and proportion manipulations only. The spatial presence is irrelevant and potentially disruptive, since the diagrams are already extremely complicated as they stand. Hence some terminological anomaly is required.

This is not yet an explanation of this use of *epi*, but we may now reformulate the problem. Archimedes' expression would have felt as strange, whatever he had chosen. He refers to a parallelepiped, an object most naturally referred to in terms of the diagram: but here, this object is a mere construct of proportion theory. Archimedes then has to import an expression from elsewhere. The remarkable thing is that he chose to import an expression from the domain of arithmetical calculations; we would have expected him to import from a nearer domain. The natural thing would have been to import the formula for a cone (27 above), to construct, on its basis, a new formula for a prism (*28 above) – as, indeed, he did in the text of *Method* 15 (see n. 96). But is this proximity perhaps the very explanation? The Greek mathematical lexicon, as I have explained, was extremely limited, and it is clear that Greek mathematicians found this useful. By being so limited, expressions were clearly demarcated from each other. The Greek mathematical expression is a territorial animal, abhorring company. Synonyms and homonyms tend to be avoided. Expressions are carefully distinguished from each other. This is a system which avoids ambiguities, and what makes it avoid ambiguities is not anything about the individual expressions, but it is about their systematic nature, their "territorial" behavior.[97] Before us is the *Sphere and Cylinder* but, despite the title, the real hero of these two books is the cone. Archimedes again

[96] They would occur, of course, wherever the prism is merely conceived, but not constructed in the diagram – e.g., in a yet unpublished text of the *Method* (coming from "Proposition 15," The Archimedes Palimpsest f. 159v. col. 2). Archimedes needs to refer to prisms that were not constructed, but can be merely envisaged as constructed on a flat diagram. Those prisms, as is obvious, are defined not by their limiting boundaries – which do not exist and are unlabeled – but by their bases (described periphrastically) and heights.

[97] I have described these features in Netz (1999) chapters 3–4.

and again manipulates cones, referring to them through expression (27) above. My guess is that *(28) was avoided because of the territorial imperative of (27), which could not bear the presence of such a close neighbor.

Whether this guess is accepted or not, we have reformulated the question. The discussion has moved on, from the logical structure of concepts, to the linguistic structure of expressions. The main fact is that Archimedes chose to import an expression from the register of calculations to the register of geometry. The innovation consists in a new configuration of registers, a new intersection of contexts.

In what context does Archimedes use the expression?

As mentioned above, the text we study is available from Eutocius' commentary on the *Sphere and Cylinder*. Let us recall the basic textual position. Archimedes promises, in the course of what we call *Sphere and Cylinder Book* II, to offer a certain lemma as an appendix to that book. This appendix was lost from the main manuscript tradition of the *Sphere and Cylinder* very early on, and certainly by Eutocius' time. Therefore Eutocius had to look for it elsewhere. Apparently, he succeeded, and he quotes his discovery in full.

If we perform the imaginary operation of reinstating this lost appendix, we have the following striking result: all the occurrences of the special *epi* we study here appear in a continuous stretch of text. They appear either in (what is now) the penultimate proposition of *Sphere and Cylinder* II, or in the appendix to that book. So this *epi* occurs only in a certain, precisely given corner of the Archimedean corpus, the very ending of *Sphere and Cylinder* II. This is a special context indeed. The second book of the *Sphere and Cylinder* is a very complex combination of proportion theory and solid geometry. Towards its end, it gets more and more complex. The alternative proof for the penultimate proposition of the book has a unique structure, effectively a theorem for which only the analysis is given. Then the appendix may be the most complex piece of mathematics in the entire corpus. Our *epi* appears as a unique expression, perhaps intentionally employed to mark a unique context. In the terms suggested in the previous chapter, we may see that the strange use of *epi* may serve to further mark a passage – to endow it with its own, distinctive aura.

The uniqueness of the expression consists in its importing from the register of calculations, to the register of geometry. Now in a way this is not unprecedented. There is another Greek tradition of geometry, which exists alongside what we consider as the main, Euclidean tradition. This is a geometry organized around calculations and not, like its Euclidean counterpart, around proofs. The main source for this type of geometry is Hero. This tradition differs from that I have concentrated on until now in many ways. For instance, Hero may easily speak of[98]

(31) The area of a figure.

He will then speak of multiplying this area "*epi*" a line. The reason Hero allows himself such a language is clear, namely, he simply chooses to ignore the phenomenon of irrationality. Since his geometry is oriented towards calculation, and not towards proof, this is a reasonable choice. The result is that Hero can identify any line with an approximate number (or, more precisely, with a combination of numbers and elementary fractions). Areas and volumes can then be calculated as approximations, on the basis of such lines.

An intriguing example of the same usage of "multiplication" of geometrical objects, in a source much more sophisticated than Hero, though no less numerical, is Ptolemy's *Almagest*[99] (the preceding context is a calculation of the radius and perimeter of the visible solar and lunar disks, measured as minutes of the great circle drawn through the sun around the center of the ecliptic, all of this within Ptolemy's theory of eclipses):

And similarly – since the radius multiplied by (*epi*) the perimeter <of the circle> makes two surfaces of the circle – of the whole surfaces: that of the solar circle is obtained as 113;6 units, and that of the lunar <circle> – as 119;32 of the same <units>.

[98] E.g., *Metrica*, Schöne (1903) 94.29. Compare 29 to *17 above!
[99] Heiberg (1898) I.514.4–8. I am grateful to Fabio Acerbi for suggesting the reference.

There is no question here of Ptolemy failing to grasp the approximate nature of the numbers used (indeed, he is certainly aware of Archimedes' own treatment of the measurement of the circle, where it is taken for granted that the measurement is an approximation). But he treats the problem as one of calculation, giving rise to actual numbers; thus the geometrical objects are treated as possessing a quantitative value, surface (*embadon* – the standard Heronian term), and they are explicitly multiplied by each other – no ambiguity about the meaning of "*epi*" here.

All of this – Hero or Ptolemy – is later than Archimedes. Yet we may assume any discussions of the measurement, in explicit numerical terms, of geometrical shapes, would have been conducted in Antiquity in this language. There is no reason to suppose Hero's discourse was original to him. In fact it clearly relates to the tradition of school arithmetic and geometry for which we have evidence in papyrus. This is all later than Archimedes – but merely for the usual reasons of the dates of papyrus survival. (I shall return to discuss this submerged tradition of non-elite mathematics in section 3.2 below.) We therefore see that Archimedes most likely did have antecedents for his practice.

Not that Hero is a perfect antecedent. There is a register-crossing here, and this is because Archimedes after all belongs to the Euclidean tradition. This is not meant as a biographical comment. I have no idea which company Archimedes would keep. But the texts we have are very clearly geared towards proof, and not towards calculation. No special numbers are mentioned in the *Sphere and Cylinder*. Rigor is the whole point of the discussion.

Archimedes' register is fundamentally different from Hero's, and the picture given above was correct. This unique stretch of text is indeed marked by this unique register-crossing. But it was useful to refine our picture of the registers. We have the Euclidean geometrical register, where no multiplication is at play at all. We have the register of calculations on numbers. We have the Heronian register, which combines the last two. And we have Archimedes' register, which at a special point, to mark it as such, opens to something quasi-Heronian. The markedness is effected precisely by this configuration of registers. The Heronian register is in general defined relative to the Euclidean one. It is a variation on the Euclidean style.

It occupies a position in the space of styles which is defined relative to the position of the Euclidean style. And therefore, when a piece which is otherwise part of the Euclidean style adopts elements of the Heronian style, this is immediately marked. Stylistic markedness rather than logical innovation: this is the effect obtained here by Archimedes.

So far I have underplayed the mathematical significance of this *epi* expression, have reduced it to an essentially stylistic innovation. But of course such innovations must ultimately have mathematical ramifications.

What indeed about the logical problem, that of the meaninglessness of multiplication in the case of (what we call) irrationals? It is clear that this should pose no problems to Archimedes. For although the linguistic practice he adopts is anomalous, the mathematical procedure itself is standard. Nothing in his formulation is such that it could not be translated into manipulations of geometrical solids. Although the language is reminiscent of that of multiplication, the practice is the same as the construction of solids. It is true that the possibility of a general parallelism between multiplication and the manipulation of lines is hinted at by Archimedes' *epi*, but this is no more than a suggestion, since this usage is limited to an isolated stretch of text, in a specialized context. Once again, we see that Archimedes opens up further possibilities, not because he is interested in exploring them himself, but, as it were, incidentally. He wishes to mark a piece of text, to endow it with its own distinctive aura. He therefore makes it different – and this difference leads on to the possibility of mathematical change.

The expression in Late Antiquity

What would happen when such an expression becomes naturalized – comes to be used within standard mathematical discourse?

In Late Antiquity this has happened – apparently separately – several times.

We may begin with Theon's commentary to the *Almagest*. Theon argues that the sphere is the greatest isoperimetric figure, comparing it in particular to the figure composed of conic surfaces introduced by Archimedes in the *First Book on the Sphere and Cylinder*.

Within this highly Archimedean terrain, Theon inserts a general statement to support his argument – and which goes beyond the Archimedean discussion (which is all confined to particular geometrical configurations). The sphere is equated with a cone, the other isoperimetric figure is equated with a pyramid, the bases are the same (as the isoperimetric assumption easily guarantees), and the height of the cone associated with the sphere is greater (the isoperimetric figure is assumed to be inscribed within the sphere). Archimedes would have stopped there, but Theon wishes to get at the reason of the conclusion. I translate Rome (1936) 378.10–16:[100]

> . . . Since indeed every cone is a third part of a cylinder having the same base and an equal height, while every pyramid is a third part of a solid having the same base and an equal height, and the cylinder is the base on (*epi*) the height, while the solid is the base on (*epi*) the height; therefore, taking the thirds, too, the said cone is then greater than the pyramid.

The measurement of solids is never defined in mainstream Greek geometry. Instead, the conditions for comparisons of magnitude are laid down by rules of proportion such as those of *Elements* XII.11 ("cones and cylinders which are under the same height are to one another as their bases"), typically proved by an indirect argument. Theon wishes to assert in a direct way the "reason" for the measurement and so he introduces, in this Archimedean context, a new thought: the geometrical object just *is* the multiplication of its measuring constituents.

This is not an isolated event. We may consider also how this expression gave rise to a more "algebraic" sense of the geometrical object, with some Late-Ancient authors criticized by Pappus (*Collection*, VII.39, adapting Jones' translation):[101]

> Our immediate predecessors have allowed themselves to admit meaning to such things, though they express nothing at all coherent when they say 'the <rectangle contained> by such and such <lines>, on the square on such and such <a line>' or 'on the <rectangle contained> by such and such <lines>'.

[100] I am grateful to Fabio Acerbi for suggesting the reference.

[101] Jones (1986) 122–3. I am once again grateful to Fabio Acerbi, who may indeed have been the first to notice the relation between this passage and the *epi* locution (indeed, he seems to be the first to realize that the *epi* locution is required for the correct translation of *Collection* VII.39).

Jones goes on in his commentary ([1986] 404) to identify the targets of Pappus' criticism. These, Jones suggests, were algebraists such as Diophantus. Diophantus, after all, had accepted expressions such as "cube-cubes" where the geometrical term "cube" can make sense only in the extended sense of generalized multiplication. Jones is probably right in identifying the milieu attacked by Pappus but we should also note that Pappus directly mentions a *geometrical* expression – "the square on the square": in other words, some authors have extended the Archimedean phrase from its original range of "area on line" to the extended sense of generalized multiplication, in this way directly algebraizing geometry.

Pappus naturally objects to this (as, indeed, his project so often is: to construct his own identity and authority by legislating on the correct procedures validated by the mathematical past).[102] He then moves on to boast his own alternative: effectively, he represents the geometrical equivalent of many multiplications not by high-dimensional objects, but by many-term compositions of ratio. Pappus is thus a witness to an avenue leading on to algebra – not a participant in this movement.

Interestingly enough, though, this very route – many-term compositions of ratio – is taken later by Eutocius (quite possibly, following upon previous Late-Ancient authors). And this once again leads – in yet a different way – to the algebraization of geometry.

With this comes the main observation with which I wish to conclude this section. Let us see how Eutocius himself goes on to use the *epi* expression, in contexts independent of the Archimedean stretch of text. In particular, in Heiberg (1915) 198.19–200.31, Eutocius introduces a lemma he has found himself. The lemma is required by Archimedes' alternative proof to *SC* II.8 – one of the Archimedean texts where the *epi* locution is used. Naturally, then, in the course of his own lemma, Eutocius often uses the *epi* expression. *This lemma, however, is in pure proportion theory*: not some unique isolated point in the outskirts of geometry, but at the

[102] I follow in this the main argument of Cuomo (2000), in her discussion of Pappus' rhetoric of the past. Incidentally, I suspect Pappus' reference to "predecessors" in the plural is a rhetorical move. He may well have in mind a single author (and in a context more like that of Book III of the *Collection* Pappus could have offered a long and detailed critique of that author, perhaps mentioning him by name). As it is, Pappus prefers to trivialize the attacked author by reducing him or her to anonymous "they."

very heart of the Euclidean geometrical discourse. Let us read this remarkable text, now from Eutocius:

Lemma to the following

Let there be four terms, A, Γ, Δ, B. I say that the ratio composed of the <rectangle contained> by A, B to the <square> on Γ, together with the ratio of B to Δ, is the same as the <rectangle contained> by A, B, on B, to the <square> on Γ, on Δ.

(a) Let the <term> K be equal to the <rectangle contained> by A, B, (b) and the <term> Λ equal to the <square> on Γ,[103] (c) and let it come to be: as B to Δ, so Λ to M; (1) therefore the ratio of K to M is composed of K to Λ – that is the <rectangle contained> by A, B to the <square> on Γ – (2) and of Λ to M (3) – that is of B to Δ. (d) So let K, having multiplied B, produce N, (e) and let Λ, having multiplied B, produce Ξ, (f) and, having multiplied Δ, <let Λ produce> O.[104] (4) Now since the <rectangle contained> by A, B is K, (5) and K, having multiplied B, has produced N, (6) therefore N is the <rectangle contained> by A, B, on B. (7) Again, since the <square> on Γ is Λ (8) and Λ, having multiplied Δ, has produced O, (9) therefore O is the <square term> on Γ, on Δ;[105] (10) so that the ratio of the <rectangle contained> by A, B, on B, to the <square> on Γ, on Δ, is the same as the <ratio> of N to O. (11) Therefore it is required to prove that the ratio of K to M is the same as the <ratio> of N to O.

(12) Now since each of K, Λ, having multiplied B, has produced, respectively, N, Ξ, (13) it is therefore: as K to Λ, so N to Ξ. (14) Again, since Λ, having multiplied each of B, Δ, has produced, respectively, Ξ, O, (15) it is therefore: as B to Δ, Ξ to O. (16) But as B to Δ, Λ to M; (17) therefore also: as Λ to M, Ξ to O. (18) Therefore K, Λ, M are in the same ratio to N, Ξ, O, taken in pairs; (19) therefore through the equality, it is also: as K to M, so N to O.[106] (20) And the ratio of K to M is the same as the <ratio> composed of the <rectangle contained> by A, B to the <square term> on Γ and of the <ratio> which B has to Δ, (21) and the ratio of N to O is the same as the <rectangle contained> by A, B, on B, to the <square term> on Γ on Δ; (22) therefore the ratio composed of the <rectangle contained> by A, B to the <square term> on Γ and of the <ratio> which B has to Δ, is the same as the <rectangle contained> by A, B, on B, to the <square term> on Γ, on Δ.

[103] Since we are dealing with "terms," two-dimensional objects can be set on the same level as one-dimensional objects: both are single-letter "terms" (i.e. governed by a masculine article). To make the reading slightly less painful, I omit the words "the <term>" from now on (as I usually omit "the <point>" and "the <line>"), but they must be understood.

[104] Anachronistically (but less anachronistically than elsewhere in Greek mathematics): $N = K^*B$, $\Xi = \Lambda^*B$, $O = \Delta^*\Lambda$.

[105] The article in the expression "the <square term> on" is masculine (for "term") instead of neuter (for "square"): a remarkable result of the semiotic eclecticism of this text, which keeps veering between general proportion theory, geometry, and calculation terms.

[106] *Elements* V.22.

And it is also clear that the <rectangle contained> by A, B, on B, is equal to the <square> on B, on A. (23) For since it is: as A to B, so the <rectangle contained> by A, B to the <square> on B (B taken as a common height),[107] (24) and if there are four proportional terms, the <rectangle contained> by the extremes is equal to the <rectangle contained> by the means,[108] (25) therefore the <rectangle contained> by A, B, on B, is equal to the <square> on B on A.

The above text is a good example of the transformation of mathematics by authors such as Eutocius. The text is limited in its mathematical ambition; it is completely governed by the terms used by classical mathematicians. Finally, the very lack of ambition, and dependence, make it startlingly original.

The mathematical ambition is limited in a very obvious way: the proposition is a lemma, i.e. conceived as no more than a tool for securing a relation Archimedes requires. This is that the composition of ratios, one with areas, another with lines, yields a ratio with "area on line" type objects. This is hardly a result at all, merely an attempt to make sense of the "area on line" expression. Quite likely, had Archimedes referred to an explicit parallelepiped, Eutocius would have felt comfortable with the argument, relying on a simple geometrical intuition that the composition of a ratio with areas with a ratio with lines yields a ratio with solids (appropriately constructed with the areas and lines involved). However, Eutocius' understanding of the situation is strictly governed by Archimedes' terms, so that he needs to make sense of the argument in terms of "area on line." This forces Eutocius into originality.

For what can Eutocius do? He has to produce some general argument that yields a result, stated in "area on line" terms. His project as a mathematical commentator is to make Archimedes' text seem to follow, everywhere, from the general mathematical assumptions an educated mathematical reader (such as Eutocius himself) would possess. In general Eutocius shows how small gaps in Archimedes' argument can be filled in by simple deductions based on Euclid's *Elements*. Essentially, this is what Eutocius does here. However, he must reach here, on the basis of Euclid's *Elements*, something that is quite different from the spirit of the *Elements*: he must reach "area

107 *Elements* VI.1. 108 *Elements* VI.16.

on line" terms. We know why Archimedes had used such terms: he wanted to enhance the special aura of his text. And now we see that the commentator, driven by the logic of his endeavor, is forced to remove such auras. Eutocius' job is to bring Archimedes' text into line with the *Elements* which means that, occasionally, Eutocius would stretch the *Elements* to make them fit Archimedes.

Specifically, Eutocius needs to do this: to operate on proportions in such a way that explicitly removes the distinction between ratios between lines, and ratios between areas. This is necessary, so that he will be able to exchange terms between the two components of the "area on line" expression. Thus the main innovation of Eutocius' text – referring throughout to the object "term" – is determined by Archimedes' expression. As noted in n. 105 above, this leads to remarkable expressions such as "(9) therefore O is the <square term> on Γ, on Δ." Since all objects are merely "terms," there are no "squares" on a line, but "square terms" (in this case, the "<square term> on Γ"). When such a square term is "on" another line (in this case Δ) the result is, directly, another term – not a geometrical object – in this case O.

There are two essential ways in which this breaks new ground relative to Classical Greek geometry. First, all objects, regardless of their dimensionality, are considered on a par (everything is a "term"). Second, objects are directly produced from each other through multiplication (O *is* the <square term> on Γ, on Δ), and are not just merely related to each other by their geometrical configuration (e.g., as one might mention that a certain square happens to have a certain line as its base). Things, in the universe of Eutocius' lemma, are defined by the multiplications and equalities that give rise to them. It is thus completely legitimate to paraphrase Step 9 by, say, $O = \Gamma^{2*}\Gamma\Delta$. Step 9 – as well as Eutocius' lemma as a whole – firmly belong to the world of algebraic equations. It is the mere formal aspect, of the lack of symbolism that still separates Eutocius from symbolic algebra. Otherwise, however, he has already left behind him the Greek geometrical conception of the mathematical object.

Archimedes' original anomaly was expressly limited to a special object, in a special context. It served its function of marking Archimedes' text. Eutocius, now, "normalizes" Archimedes' text by reaching it from Euclid's general proportion theory. To do this, however, is to extend Archimedes' anomaly to all objects, and to a context much more general than Archimedes had intended. This extension, in itself, immediately leads to the algebraization of the text: since all objects are to be defined in terms of relations equivalent to the "area on line" expression, everything is given in terms of multiplications and equations – which is already within algebra.

How can Eutocius feel justified in being so original? Paradoxically, this is because of his deep respect to the past. By the time Eutocius comes to write his commentary on Archimedes, Greek mathematics has become, for him, a certain *canon*. Archimedes could convey meanings to his readers, by positioning parts of his texts in different ways relative to different registers. The meaning would be conveyed by the relative positions of those registers. But the very passage of time has created a new configuration, in which the very fact that a text is Archimedean endows it with a canonic significance. The space of registers has been transformed. Most important to notice, this is not a historical accident, but a necessity. Wherever there is canon-formation, where there are "classics," the very classicality of the classics gives them a position in the space of registers which they never could have occupied while still "alive," in their own original context. Since part of the meaning of texts is conveyed by their use of registers, it follows that some of the meanings inside a corpus will necessarily alter simply by the fact that this canon was canonized. In our case, the canonization of Archimedes has meant that the Archimedean *epi* lost its marginality. What Archimedes had intended as a local aura created by a stylistic effect has become for Eutocius a natural practice, sanctioned by the authority of Archimedes. So this is the general rule: canonization flattens the hierarchization that was intended to hold between the registers of the original. Can we really sense the deep bathos between Shakespeare's poetical soliloquies and his rude jokes? All are equally Shakespearean to us now. In the case of Archimedes and Eutocius – of Greek mathematics and its later readers – this flattening of registers had subtle conceptual consequences. As later readers broke down the borders between registers, geometry became algebraized.

2.6 The problem in the world of Eutocius

Nothing in the text of Eutocius suggests any new conceptualization of mathematics. The objects are geometrical, and they are treated, generally speaking, with the tools available from Euclid's *Elements*. In fact, Eutocius' original mathematics tends to be very elementary in character, so that he has little occasion to refer, himself, to conic sections. His mathematics is, generally speaking, that of the simplest plane geometry and proportion theory.

This is of course as would be expected. Eutocius did not set out to write about geometry; he set out to write about Archimedes' geometrical treatise. Several recent studies in the practices of mathematics point towards an explanation of this. Cuomo (2000) is a study of Pappus' *Collection* (fourth century AD): it is shown how Pappus builds his intellectual credentials by reference to past authorities. What makes Pappus an important intellectual figure is his mastery over a mathematical canon and not his own intellectual originality. The polemic tone of early Greek mathematics is still recognizable in Pappus, but now the contemporaries (as well as, sometimes, the ancients themselves) are criticized for failing to live up to the inherited canon. The goal of science, then, is redefined: works should fit a certain canonical pattern.

Mansfeld (1994) discussed a central form of the construction of such past canons, in Late-Ancient culture in general – which he then applied to mathematics, in Mansfeld (1998). A major genre in that period was the "Introduction," which had acquired a relatively fixed format. More and more, authors were concerned not only with presenting new, original ideas, but in arranging works inherited from the past. The introduction makes an argument for such an arrangement. Typically, the introduced author is put in the context of other works by himself, so that, for instance, the introduction suggests the preferred order of reading that author. He is also put in the context of other works in the same genre, explaining the place of the author in the canon.

We see then that a large part of intellectual activity in Late Antiquity was involved not with writing about things, but in writing about books. This is writing which is essentially dependent upon some previous writing – what I call a *deuteronomic* text. Deuteronomic texts are fundamental to the cultures of the Mediterranean codex, from at least the third century AD to perhaps as late as the fifteenth century. Throughout this period, a large part of intellectual life is engaged with such typical deuteronomical activities as translation, commentary, epitomizing and completion of earlier works.

In Netz (1998) I have suggested several consequences of deuteronomy in the history of mathematics which I shall now sum up briefly. In a deuteronomic setting, the most natural thing to do is to fill in gaps in original arguments. Thus new texts are formed which are more "tight" in their argumentation, often to the point of pedanticism (this is obvious in the glosses added by Eutocius in the course of Archimedes' argument for the limits of solubility; it appears – judging from the Arabic version – that Eutocius did something similar with the text of Diocles, as well). Another natural thing is to add in the discussion of further *cases* which, I argue, was done by Eutocius in his commentary to Archimedes' discussion of the limits of solubility.

Deuteronomic authors often standardize the text: e.g., it is made to fit some established structure of presentation, or by making the structure more explicit. (For instance, many of the QEDs we have at the end of mathematical propositions – as well as the proposition numbering we have at their beginning – may come from this kind of activity.) Deuteronomic authors, in the course of their discussions, often introduce a second-order terminology for classifying mathematical texts (the well-known classification into kinds of problems – planar, solid, and linear – seems to come from such a context).[109] Another way of standardizing the text is by making its various parts cohere with each other logically (e.g., by pointing out that certain propositions can be derived from others, which the original author might have left implicit; indeed, in some cases, deuteronomic authors provide links between propositions that, originally, were each proved separately). In general, deuteronomic authors tend to make the works they study homogeneous: for instance,

[109] The *locus classicus* for this is Pappus, book III; see the discussions in Knorr (1986) 341 ff., as well as Cuomo (2000) chapter 4 – which is much more sensitive to the historical setting from which such schemes derive (in particular, Cuomo discusses in detail Pappus' criticism of a purported solution, for failing to recognize the nature of the problem as established by tradition).

similar letters are used in similar contexts, and a general attention to "correctness" (understood by reference to the established canon) permeates all work. Many more examples can be given, but the overall nature of deuteronomic texts is clear: they tend to systematize. When working on a mathematical issue, the goal is to bring it to its proper place inside Mathematics with a capital M – the Mathematics which is an ideal book-type object, containing all the cases and all the details in perfect order. Such is deuteronomic culture. It may sometimes, indeed, be merely pedantic; add in genius, however, and you may get important works of mathematics, as we shall see in the next chapter.

Eutocius is, in a sense, "merely pedantic." He has little occasion to make new conceptual breakthroughs. He also has little motivation to do so. His goal was merely to explicate Archimedes' text, to explain how the results are obtained and to fill in gaps in the argument. He is thus completely dependent on Archimedes' text: obviously, a *deuteronomic* author.

Still, we saw that, returning to the material of Archimedes' problem, Eutocius produced, on two separate occasions, a text alien to Archimedes' own approach. In section 2.4, we saw how Eutocius treated the conic section almost as if it were a set of points defined by functional relations (and not as a spatial unit, defined by geometrical properties). In section 2.6, we saw how Eutocius treated all geometrical objects as mere quantitative magnitudes, in principle producible from each other by multiplication (and not, as Archimedes had them, as individual geometrical objects, each defined separately by its qualitative features).

We see a certain affinity between the two transformations: Eutocius, somehow, levels the geometrical universe: things become more closely internally related (such as the points on the curve) and externally related (such as the various magnitudes, reciprocally producible by multiplication). The same, indeed, happens with Theon (who "levels" the geometrical and arithmetical worlds so that bases are multiplied by heights) and with the authors attacked by Pappus (who "level" the geometrical and arithmetical world by allowing many-dimensional multiplications).

Why should this happen? Let us now see in detail the basis of Eutocius' mathematics in his deuteronomic practice.

Consider first the first transformation, discussed in section 2.4. On two occasions, from Archimedes and from Diocles, we saw Eutocius intervening in the text in relatively minimal ways: adding a very simple case to Archimedes' treatment of the limits of solubility, adding the words "and vice versa" into the text of Diocles. In both cases, then, Eutocius had little mathematical ambition: he simply added something that, in principle, was already there. The situation, in both cases, was as follows. Archimedes and Diocles had shown a general property, but they had mentioned it only for one case, the one present in the diagram. This, then, was typical for Classical Greek mathematics. The mathematical object was approached not as a general entity, defined by some abstract properties, but as a specific individual residing inside the diagram. That the mathematical argument held generally, apart from the diagram, was seen to be the case through the implied repeatability of the same argument to all similar diagrams. Such a repeatability, however, was merely implied by the text which was limited, throughout, to the individual diagram at hand. Such is Klein's interpretation of Greek mathematics (for which I have also argued in chapter 6 of Netz [1999]). We can now see this principle in terms of Classical mathematical practice, as described in the preceding chapter. Of course Classical Greek mathematicians aimed to have their arguments developed in terms of the particular diagram: they aimed to endow their arguments with a particular aura, independent of any other argument or diagram. The possible generalization of the argument had to be left implicit since, otherwise, the individual properties of the diagram would be seen as incidental to the argument. Hence the claim was made for a unique diagram, uniquely drawn and labeled.

In both texts of Archimedes and Diocles, we see this in practice. The argument, quite simply, is developed for only one case: the diagram bifurcates, in principle, into two symmetrical cases, and both authors, Archimedes and Diocles, chose to concentrate on one of the two alone.

Eutocius, however, stands back from the particular text he has available to him from Archimedes and Diocles, and immerses it in a wider context. He sees the particular diagram as a single case of a family of similar diagrams. He asks himself such questions as:

what would the diagram of Archimedes' treatment of the limits of solubility look like, were the point to be taken left of the maximum? What would the diagram of Diocles' solution look like, were the ellipse to extend to the left, and not to the right of the base line? It is clear that such questions can have deep consequences – though they might have no deep motivations.

Such questions are natural to ask in a commentary: to Eutocius, there is no longer the need to isolate the individual diagram. On the contrary, he approaches it from the outside and wishes to understand it, and so puts it in the context of other possible, similar diagrams. His writing is determined by the Classical texts, but what he does with them is to put them inside a context – which is precisely what they had avoided doing. Archimedes and Diocles aimed at the individual aura; Eutocius aims at contextualization, which is the removal of aura. Hence, Eutocius' mathematics has concepts that are different from those of Archimedes and Diocles, and are different in a well-defined way: Archimedes and Diocles perceive their objects as standing apart from each other, Eutocius sees them as continuously dependent upon each other.

A similar story can be told, at a more global level, for Eutocius' second transformation, discussed in section 2.5. Once again, Eutocius does not reflect on concepts: he simply appropriates an Archimedean expression. But this immediately implies that the same expression shall be used in a context different from that intended by Archimedes. Archimedes' expression, "the area on the line," was designated as an intentionally strange expression, enhancing the distinctive aura of a distinctive text (appropriately, one dealing with a proportion of areas and lines). Archimedes' text was meant to stand apart from, say, Euclid's proportion theory. As a commentator, Eutocius explicates everything in terms of the mathematical education available, in principle, to his contemporaries. Everything, therefore, is brought to the level of Euclid's *Elements*. This includes Archimedes' expression "the area on the line": Eutocius needs to derive properties of that expression, in the terms of general Euclidean proportion theory. This brings him, as we saw above, to even more remarkable expressions, such as "O is the <square term> on Γ, on Δ." All objects are reciprocally producible in terms of operations such as "square" and "on" (i.e.

multiplication). This is very different from Archimedes' usage, where the strange expression can be used in very limited circumstances, so that it is felt as an anomaly, not touching upon the basic understanding of mathematical objects in general.

The reason for Eutocius' transformation, in this case, is especially interesting. We have seen already Eutocius' need, as a deuteronomic author, to bring things together. As a commentator, he puts things into contexts – whether putting a diagram in the context of other possible diagrams (as we saw above) or putting a special expression in the context of Euclidean proportion theory (as we see here). But notice that in this particular case, Eutocius' need is more urgent, because the original *resists* being put in context. Archimedes' text is expressly anomalous: it is characterized by certain dualities and tensions. As a commentator, Eutocius aims, it seems, to remove such tensions – to harmonize whatever he has available to him from the Classical tradition. To Eutocius, there is a value in the very homogeneity of the tradition. But we have seen that ancient authors aimed at being distinct from each other, so as to endow their works with their distinct auras. As a commentator, then, Eutocius simply *had* to remove such auras. The project of the commentator was, in a certain way, diametrically opposed to that of the Classical Greek mathematician. In this case, Archimedes wanted to create a special object, subsisting in a unique body of mathematics – "the area on the line." Eutocius, without even realizing this, removed the uniqueness of the object by subsuming it within general proportion theory. And in order to do this, he had to bring all objects together, reduced to mere "terms" that are tied by such quantitative relations as "square" and "multiplication."

There are thus two forces that contribute to the dynamics of change, leading from Classical Greek mathematics to its later transformations. First, there were changes in the practices of mathematics; second, there were inherent tensions inside the practice of Classical Greek mathematics itself, that would serve as basis for such later transformations. We shall return to discuss such forces in the conclusion.

Let us survey the ground covered so far. In chapter one we saw how the practice of Classical Greek mathematics, aiming at the individual aura of individual arguments, naturally gave rise to mathematical objects that are defined by the particular qualitative features of local configurations. In this chapter we have seen how the practice of a deuteronomic text – Eutocius' commentary to Archimedes – naturally gave rise to mathematical objects that are tied together by functional relations, and participate in equations.

At the level of Eutocius' mathematics, however, these are no more than hints of possible mathematical change. Eutocius simply says little, as a mathematician. In particular, he did not try to offer his own solution to Archimedes' problem. The real question is this: what happens, when a solution is offered in the context of deuteronomic mathematics? This question is answered in the Arabic world.

3

FROM ARCHIMEDES TO KHAYYAM

In this chapter we concentrate on the fate of Archimedes' problem in one eminent work of Arabic science: Omar Khayyam's *Algebra* (eleventh to twelfth centuries). (This is, of course, the same Omar Khayyam famous for his Persian poetry; here we concentrate on his science.)

As we shall see below, this decision to focus on Khayyam is to a certain extent arbitrary: the problem had a significant history in the Arabic world before and after Khayyam. He does occupy a special position in the history of the problem. Our knowledge of Arabic treatments prior to him is in some cases derived from him alone (much as we know of early Greek treatments of the problem through the work of Eutocius). And while the later history of the problem adds much that is mathematically valuable, we can usefully end our survey with Khayyam. With him, as we shall see, the route from problems to equations is largely completed. It is also helpful to compare like with like: and it is therefore appropriate to have our survey – begun with the genius of Archimedes – end with the genius of Khayyam.

Our goal in this chapter, then, is to show that Khayyam's mathematics already differs essentially from Archimedes'. This should be a deep conceptual divide, along the lines suggested by Klein and Unguru. We also need to show the historical basis for this divide, in terms of changes in the practice of mathematics from the world of Archimedes to the world of Khayyam. Throughout, we shall argue for a continuity between the originality of the world of Khayyam, and that of the world of Eutocius. Similar transformations characterize the science of Greek Late Antiquity, and that of the Arabic world. In the Arabic world, those transformations are much more thoroughgoing and ambitious, resulting in a new kind of science.

Section 3.1 surveys in outline the history of Archimedes' problem in the Arab world up to Khayyam. Section 3.2 adds another essential background to the work of Khayyam: the new Arabic science of "Algebra," as inaugurated by Al-Khwarizmi. In section 3.3 we put Khayyam's solution to the problem in the context of Khayyam's *Algebra*, and in section 3.4 we read Khayyam's solution itself. This is followed by two comparisons: in section 3.5, we compare Khayyam's solution to that of Archimedes. In section 3.6, we compare Khayyam's polemical style to the polemical style of Greek authors such as Dionysodorus and Diocles (we also glance briefly at the advancement of Khayyam's own work by his follower, Sharraf Al-Din Al-Tusi). In both sections 3.5 and 3.6 we find that Khayyam – like Eutocius before him – differs from Classical Greek mathematicians in an essential way: Classical Greek mathematicians aimed to set their work apart from that of their peers, Khayyam and Eutocius primarily aimed to put their work in context. In section 3.7, the conclusion, we argue that this results from Khayyam's cultural practices which, like those of Eutocius, were *deuteronomic* – he was the author of texts essentially dependent upon previous texts. This, in turn, accounts for the originality of Khayyam: the construction of the modern "equation" and, indeed, modern algebra.

The richness of the Arabic material prevents the same kind of detailed treatment we employed for the Greek material. In this chapter, I look into the detail of one representative author – Khayyam – and survey more briefly the rest of the Arabic tradition of the problem, noting, throughout, where Arab mathematics breaks away from the Hellenistic tradition, where it is continuous with Late Ancient practice, and where – and why – it begins to suggest our own modern algebra.

3.1 Archimedes' problem in the Arab world

Let us first recall the fate of Archimedes' problem *prior* to the coming of the Arab world. Between the time of Archimedes and that of Eutocius, Archimedes' solution to the problem of the proportion with areas and lines was effectively unknown. That is: it survived only in a manuscript tradition separate from that of the much better known books *On Sphere and Cylinder*. Thus readers of the *Second Book on the Sphere and Cylinder* could very easily be led to believe that Archimedes had merely promised to offer a certain solution, which in fact he had left unproved. This was the view of Dionysodorus and of Diocles. They did not have access to catalogues containing all known specimens of Archimedes' work. They had merely read the books *On the Sphere and Cylinder*, and could not know that, in some obscure volume, in some obscure library, one could also find the missing key piece of the text. How many times did we, modern scholars – with all our library catalogues and bibliographies – commit similar errors of omission!

Eutocius' commentary to the works of Archimedes did not change this situation in a fundamental way. The same bifurcation still could exist, now between the tradition of Archimedes' works, and that of Eutocius'. On only one known occasion, a compilation of works by Archimedes was combined with a compilation of Eutocius' commentaries to Archimedes. This special combination was the prototype of an important Greek manuscript, copied in the ninth century, and called by Heiberg "Codex A." This codex was the foundation for almost all knowledge of Archimedes in Western Europe, from the thirteenth century to the nineteenth century. Thus, from the perspective of Western Europe, Archimedes and Eutocius' commentaries seem to be tightly related: but this is an illusion caused by the accident of a single codex. In all other lines of transmission – Greek, Latin, Arabic, or Hebrew – Archimedes and Eutocius are only slightly related.

This is true, in particular, for the Arabic tradition. Archimedes' books *On Sphere and Cylinder* were translated twice into Arabic, once, indirectly, via Syriac, and once directly from the Greek (both, apparently, made as early as the ninth century).[1] Eutocius' commentary, in whole or in part, was translated apparently not later than the tenth century. While several extant Arabic manuscripts have both of Archimedes' books *On Sphere and the Cylinder*,

[1] For the Greek and Latin tradition of Archimedes, see Heiberg (1915), Clagett (1964–84). For details on the Arabic and Hebrew tradition of Archimedes' books *On Sphere and the Cylinder*, as well as Eutocius' commentary, see Lorch (1989), Sezgin (1974).

unabridged, and often together with some further works by Archimedes (or ascribed to him), the manuscript tradition for Eutocius is much thinner. Only one manuscript has a complete text of Eutocius, namely (almost) all the commentary to the *First Book On the Sphere and Cylinder* (perhaps from the early fourteenth century, this manuscript – Bodl. Heb. D. 4 – is a further translation, from Arabic into Hebrew). The rest are selections, only two of which include Archimedes' solution to the problem of the proportion with areas and lines. These selections sometimes surface together with Archimedes' books, sometimes separately from them. Apparently, only one extant manuscript – Istanbul, Fatih 3414 (thirteenth century) has both Archimedes' books *On the Sphere and Cylinder*, as well as a selection from Eutocius including Archimedes' solution, all in the same codex.

In short, given this basic situation, we should expect many Arabic mathematicians to be exactly in the same position as Dionysodorus and Diocles, and to offer their own solution to Archimedes' problem as a completion of Archimedes' text. This list is distinguished: Al-Mahani (ninth century), Al-Khazin (tenth century), Al-Quhi (tenth century), Abu'l-Jud (tenth to eleventh centuries), Ibn Al-Haytham (tenth to eleventh centuries), Omar Khayyam (eleventh to twelfth centuries) and finally Sharraf Al-Din Al-Tusi (twelfth century). (We can see then that Archimedes' problem flourished in the second century BC, with the works of Dionysodorus and Diocles, was briefly revived in the sixth century AD by Eutocius, and then reached its period of most intensive study in the ninth to the twelfth centuries. This curve is typical to the history of early Mediterranean mathematics.)

The reason for the interest of Arabic mathematicians in the problem is similar, in outline, to that of the interest of Dionysodorus and Diocles. Considered in more detail, one finds a subtle difference. Whereas the interest of Classical Greek mathematicians is to criticize Archimedes (or other previous mathematicians), the interest of Arabic mathematicians is, rather, to complete him. A good example for this is Al-Quhi. As mentioned above, Al-Quhi was among the Arabic mathematicians who discussed the problem of the proportion with areas and lines. Alongside that, Al-Quhi offered a further completion of Archimedes' *Second Book on the Sphere and Cylinder*, where Archimedes, among other things, solves the following problems:

- To find a segment of a sphere similar to a given segment, and equal to another given segment (Proposition 5).
- To find a segment of a sphere similar to a given segment, its surface equal to a surface of a given segment (Proposition 6).

To this, in the tenth century, Al-Quhi has added the following problem:

- To find a segment of a sphere equal to a given segment, its surface equal to a surface of a given segment.

This last problem is, in fact, even more complicated than anything Archimedes attempted, and its elegant solution by Al-Quhi is typical of the way in which Arabic mathematics fully matched its Greek origins.[2] What I want to stress here is the type of motivation Arab mathematicians had. Al-Quhi faces Classical results – in this case Archimedes' two problems. He then notices that they can be put inside a context: various problems defined by given surfaces and volumes, similarity and equality. Once Al-Quhi's problem is added, the context is complete: no more problems can be added.[3] As Al-Quhi had stated explicitly, his goal was the completion of Archimedes' second book.[4] Clearly Al-Quhi, just like his Greek predecessors, wishes to make a name for himself, by comparing himself, favourably, with the greatest mathematician, Archimedes. But the route he chooses – completion – differs essentially from the Greek route of criticism. Criticism implies that one returns to stand in Archimedes' position, to rework Archimedes' original solution – and we have seen this in the case of Dionysodorus and Diocles. Completion demands that one stand back from Archimedes' position and consider it in context. We also note that criticism does not

[2] For discussion and partial translation, see Woepcke (1851) 103–14. One will always admire Woepcke's combination of historical and mathematical good sense. That, over the last century and a half, Arabic mathematics still has not received the full historical attention it deserves, is a scandal.

[3] One could imagine one further problem, to find a segment of a sphere equal to one segment, similar to another, and its surface equal to the surface of a third. This however is only soluble in trivial cases, since the combination of the two equalities – with a volume, and with a surface – already determines the shape of the segment.

[4] Woepcke (1851) 104.

make any positive use of Archimedes' original text that, instead, it aims to supersede; while completion essentially depends on the original text to which it aims to make a supplement. Completion, unlike criticism, is part of a *deuteronomic* project: the creation of new texts that are essentially dependent upon the Classical texts.

It is in this spirit, then, that Arab mathematicians had approached the more obvious lacuna in the *Second Book on the Sphere and Cylinder* – the absence of a solution to the problem of the proportion with areas and lines. The goal was to patch the gap. Hence, there was no attempt to reconsider the problem from scratch, as Dionysodorus and Diocles did. It is typical that Al-Mahani – apparently the first Arabic author on the subject – reached it in the course of his *commentary* on Archimedes. His project, then, was directly equivalent to that of Eutocius. This commentary is not extant, but we can conclude, on the basis of later testimony, that Al-Mahani was incapable of filling in the gap. (Apparently, then, he himself did not find Eutocius' commentary.) Instead, he took the problem from the form Archimedes had left it in – a proportion with areas and lines – and transformed this to a contemporary form. Al-Mahani lived not long after Al-Khwarizmi, who had shown how to present geometrical relations of areas and lines in terms of multiplications and additions. Translating Archimedes' problem into the terms of multiplications and additions, then, Al-Mahani reached, in all probability, something like:

> A cube, together with a given magnitude, equals a square multiplied by another given magnitude.

We have discussed in section 1.3 above the mathematical basis of this reformulation. It is clearly a major step on the route from problems to equations, and we shall stop to consider its significance in the following section. But note, meanwhile, how minimal were Al-Mahani's goals. He did not aim to change Archimedes' problem, but to solve it. He was incapable of doing so, however, and therefore he did the one thing he could do: he reconsidered the problem. In all likelihood, had Al-Mahani been able to supply Archimedes' lost solution, he would not have transformed Archimedes' extant statement. Given his position, however, a natural route forward would be to reformulate this reduction in the canonical terms of Al-Mahani's time. All Al-Mahani ever did was to put Archimedes' formulation in the context of the wider set of expressions available to him, Al-Mahani, but not to Archimedes. As we shall see, no Arab mathematician attempted to reformulate the problem in terms that differ markedly from Archimedes. All Arab formulations are conspicuously equivalent to those by Archimedes. In other words, there is no seeking of a distinctive aura, setting one's problem apart from other, established problems.

At any rate, Al-Mahani did not try to offer a solution to the problem, even in the seemingly tractable form he had obtained. Abu'l-Jud, apparently, had made the attempt. He had taken the problem in the form bequeathed by Al-Mahani. Once again, Abu'l-Jud's solution itself is lost: this time, however, we have some information about it, as Khayyam takes the trouble to show that this solution was, in fact, *false*. It appears that both Khayyam's formulation of the problem, as well as his solution, were directly comparable to Abu'l-Jud, with the difference that Khayyam, unlike Abu'l-Jud, realized that, given certain conditions, two conic sections intersect, giving rise to two solutions (Abu'l-Jud believed they were tangents there, with only one solution allowed) and that, given other conditions, at least one solution was possible (Abu'l-Jud thought those conditions made the problem impossible). We shall return to discuss this in detail in section 3.6 below, as Khayyam's criticism of Abu'l-Jud provides us with an opportunity to study the nature of polemic in the Arab world. We note meanwhile that (a) Abu'l-Jud took the problem in the form bequeathed to him by the tradition, (b) he considered it in terms of the cases to which it gives rise.[5]

Even prior to Abu'l-Jud – though, apparently, the two might have worked independently – Al-Khazin had already offered a correct solution to Archimedes' problem. Once again, our information is derived from Khayyam alone and, in this case, we know even less: merely that Khayyam considered Al-Khazin's solution to be sound. Since Khayyam does not mention any improvement he has made over Al-Khazin for this particular problem (rather, as we shall see below, his pride is in having systematized all problems of the same type), the likeliest conclusion is that Al-Khazin had obtained,

[5] Our information on Abu'l-Jud's solution derives from Khayyam alone.

effectively, the same solution as we have from Khayyam himself. It appears, then, that just like Abu'l-Jud after him, Al-Khazin (a) solved the problem in Al-Mahani's terms, and that (b) he offered a consideration of the cases. (Khayyam's contrasts with Al-Mahani, who expressed the problem, to Al-Khazin, who solved it, suggesting conclusion (a); further, without a consideration of the cases, Khayyam would hardly have considered Al-Khazin's treatment sound.)

Abu'l-Jud might not have been aware of Al-Khazin's previous work on the problem. It is possible that Khayyam, in his turn, was unaware of Al-Quhi's and Ibn Al-Haytham's previous works.

Once again, the two approaches – by Al-Quhi and by Ibn Al-Haytham – seem to have been independent of each other. In fact, Ibn Al-Haytham seems to have been independent of the entire Arabic tradition, in that he did not start from Al-Mahani's formulation, but returned to Archimedes' own formulation of the problem. Thus his project is that of completion in the most limited sense, namely, to fill in the gap in Archimedes' reasoning, in precisely the terms stated by Archimedes. It is interesting that Ibn Al-Haytham explicitly states the task in such terms. In paraphrase, Al-Haytham explained that Archimedes' solution of the problem was correct but that, in order to state the solution of the problem with areas and lines, Archimedes had to employ conic sections. However – explains Ibn Al-Haytham – Archimedes preferred not to do so, so as not to encroach on the style of the *Second Book on the Sphere and Cylinder* (which, in its extant form, does not mention conic sections). Thus, to justify Archimedes (that is, not at all to criticize him!) Ibn Al-Haytham went on to supply the missing argument.[6] (I note in passing that Ibn Al-Haytham's historical guess as to why Archimedes had removed the missing proof to an appendix is quite convincing.) We see here a clear example of the deuteronomic project of Arab mathematics, completing previous works rather than criticizing them. Given this background, it is obvious why it would be best, for Al-Haytham, to solve Archimedes' problem in *precisely* the terms stated by Archimedes. The solution itself is effectively equivalent to that quoted by Eutocius, so

[6] Woepcke (1851) 91.

that one wonders whether Ibn Al-Haytham, perhaps, did not have access to Eutocius or some other version ultimately derived from him. At any rate, Ibn Al-Haytham then goes on to offer a second solution to the same problem, once again considered in the same original terms stated by Archimedes, though this time constructing a mechanism that conserves the geometrical relations required by the solution.[7] This, then, is a strictly spatial understanding of the problem. This particular set of solutions from Ibn Al-Haytham is an example for Arab mathematics that is clearly concerned with problems, not with equations. As we shall see in sections 3.3–5 below, when we discuss in detail Khayyam's solution, it is not as if Arab mathematics crossed any conceptual Rubicon, so that the objects of mathematics were no longer considered spatially and geometrically. Conceptually, one notes a continuity with Greek mathematics. The historical transformation had to do not with concepts, but with practices. Ibn Al-Haytham, however, gives a fine example of the practice of Arab mathematics, deuteronomic rather than aura-seeking. As we shall see, this change in practice could occasionally lead to what we may call the study of equations.

Al-Quhi, perhaps, came close to this. We have mentioned above Al-Quhi's completion of Archimedes' problems on the segments of the sphere, as an example of the deuteronomic project of Arab mathematics. Al-Quhi's contribution to Archimedes' problem went beyond that. We do not know if Al-Quhi had offered, independently of other authors, his own solution to Archimedes' problem in the form given it by Al-Mahani (he was certainly aware of Al-Mahani himself). But we do know that he considered the question of the limits of solubility, and had stated it differently from the form quoted by Eutocius. In the text reported by Eutocius the limit is stated as follows:

> BE being twice EA, the <square> on BE on EA is <the> greatest of all <magnitudes> similarly taken on BA.

Al-Quhi's terms (reached, probably, independently from Eutocius' text) are:

[7] Woepcke (1851) 92–6.

the line C [that is, the side of the given square in the terms of the problem stated by Archimedes] is not greater than . . . [I skip Al-Quhi's derivation] . . . that is, the line on which is constructed four ninths of the third of the cube AB.[8]

Woepcke, not without reason, transcribes this as:

$$C \leq \sqrt[3]{(4/27)AB^3}$$

which, as we will recall, was Heath's statement of the limit. In other words, Al-Quhi seems to reach something that is rather like the algebraic reading of Greek mathematics. This is obtained by a very simple transformation, essentially that already effected by Al-Mahani: the reformulation of Archimedes' problem, in the terms of Al-Khwarizmi (that is, representing geometrical relations as multiplications and additions). Clearly, then, this very transformation is of great historical significance. In the next section, I briefly discuss the historical origins of Al-Khwarizmi's project.

3.2 A note on Al-Khwarizmi's algebra

Al-Khwarizmi, one of the earliest and most influential of Arabic mathematicians, was a court scholar: a member of Bayt al-Hikma, the "House of Wisdom" founded in Baghdad by the caliph Al-Mamun. During that reign (AD 813–33) there seems to have been an extraordinary period of activity in the "House of Wisdom," setting the stage for all later Arabic science. It was then that Al-Khwarizmi produced his book *On the Art of Al-Jabr wa l-Mukabala*. The book – like much else produced at that time and place – had an enormous influence among future authors, in Arabic and in other Mediterranean languages. As is well known, the very word "algebra" derives from the title; let us now consider the relation between the contents of the book and algebra itself.

The work is a miscellany of problems of calculation, dealing with all that "men constantly require in cases of inheritance, legacies, partition, lawsuits, and trade, and in all their dealings with one another, or where the measuring of lands, the digging of canals, geometrical computations [i.e. land-measurement] and

[8] Woepcke (1851) 96–103.

other objects of various sorts and kinds are required."[9] Most of the work is at this level of particular detail, with the majority of the problems related to the study of divisions in legacies. The very beginning of the treatise gained most scholarly attention, as it approximates most closely our sense of "algebra."

Al-Khwarizmi starts by mentioning three kinds of objects: "quantity" (roughly, a square of an unknown in an equation), "root" (roughly, an unknown in an equation),[10] and "number" (roughly, a constant in an equation). Note that Al-Khwarizmi does not obtain any consistency in his terms. He always seems to operate by having, not far from the more abstract statement, a more natural decoding in terms of practical calculations, so that, for instance, instead of "numbers" he might occasionally speak of "Dirhams" (a standard coinage unit).

After mentioning the three types of objects, Al-Khwarizmi moves to a survey of the kinds of problems that may be stated and solved with these three, the first being "quantity equal to roots." Al-Khwarizmi does not dwell on this general form, but immediately mentions an actual value of a problem arising within it, "a quantity equals five of its roots" which – Al-Khwarizmi simply states – means that the root is five, and the quantity is twenty-five. In this first problem there is only one quantity, which makes it truly trivial. The next problem, "a third of a quantity equals four roots" is already more complicated though, once again, Al-Khwarizmi merely states the answer.

In some more complicated problems Al-Khwarizmi does go through a process yielding the result, e.g., when he deals with "quantities and roots equal to numbers." The first problem is "quantity and ten roots equal to thirty-nine Dirhams." There you are told to divide the (number of the) roots by two, to get five. This,

[9] Rosen (1969) 3. For the text of Al-Khwarizmi, confer also Musharrafa and Ahmad (1939), as well as the useful editions of the Latin translations, in particular Hughes (1989), Karpinski (1915).

[10] Note that for Al-Khwarizmi it is sometimes natural to consider the second power as the baseline (the *quantity*, simpliciter), and consider the first power as its *root*, whereas we usually take the first power as the baseline (x, simpliciter). This is most natural if we envisage Al-Khwarizmi's problems at the start of the book as interpreted in terms of field measurements, where the goal of the discussion is the value of the two-dimensional field itself. It should be noted that the term "quantity" might have originally meant "property."

multiplied by itself, yields twenty-five. This, added to the given thirty-nine, yields sixty-four. The root of that is eight which, when the half of the (number of the) roots is subtracted from it, yields three. This is the root and the quantity, therefore, is nine. This route of solution is neither generalized nor justified by Al-Khwarizmi (we shall soon consider a justification for that).

In all, Al-Khwarizmi lists the following kinds of problems:

- Quantities equal to roots,
- Quantities equal to numbers,
- Roots equal to numbers,
- Quantities and roots equal to numbers,
- Quantities and numbers equal to roots,
- Roots and quantities equal to numbers.

Having discussed all six kinds by such particular examples and particular, unjustified solutions, Al-Khwarizmi states (once again, without argument) that these exhaust the kinds of problems to be investigated.

(It should be noted, incidentally, that problems might sometimes fall outside those canonical forms, e.g., when we have "A [number of] quantities equals a [number of] roots minus a [number of] quantities." In such cases the problem might be "completed" by transferring the "minus a [number of] quantities" expression, as a positive value, to the expression "a [number of] quantities," so that we have the form "A number of quantities equals a [number of] roots." This operation is known as "Al-Jabr" and, together with a similar operation, "Al-Mukabala," it gives its name to the treatise and eventually to algebra.)

Following upon this survey of types of problems and their solutions, Al-Khwarizmi offers a geometrical interlude before moving on to other types of problems of calculation. In this interlude, he offers "reasons" for some of the solutions mentioned earlier. The first such treatment is of the problem we quoted above, "quantity and ten roots equal to thirty-nine Dirhams," for which Al-Khwarizmi offers two separate geometrical reasons. We shall now paraphrase the second and more simple. (fig. 13). Position the quantity as the square AB. Attach to it two rectangles, G and D, so that both their lengths (that is, the side apart from the side of

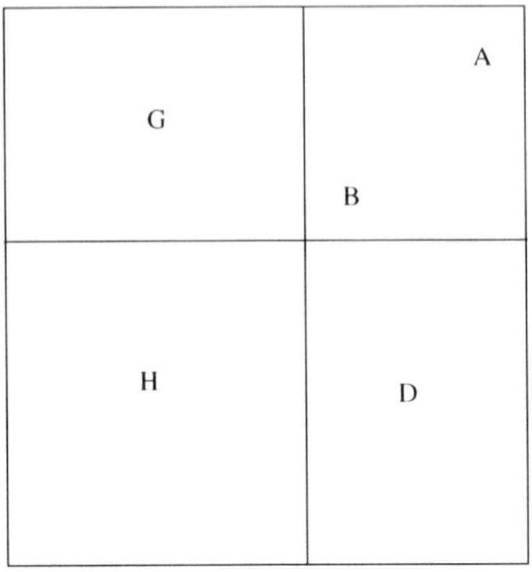

Figure 13

the square) are five. [It now follows, then, that the value of each of these two rectangles is five times the side of the square, or five times the root, that is, both together are equal to ten times the root.] The remaining square H is equal to twenty-five (each of its sides is five). Thus the entire square composed of the four figures equal the given square, ten roots, and twenty-five. We know, from the terms of the problem, that the given square with ten of its roots are equal to thirty-nine; add thirty-nine, and the entire square is equal to sixty-four. Hence its side is equal to eight and, when subtracting the five which was added, we get three as the side of the original given square.

Following this geometrical interlude, Al-Khwarizmi moves on to more detail about calculations, which we may ignore here. The fundamental features of Arabic algebra are already in place, and we may begin to consider the questions that interest us: where are we on the trajectory from problems to equations, and what, in the mathematical practice of Al-Khwarizmi, explains his position?

The two main features of the work, from our point of view, are systematization (the attempt to arrange problems into six types), and the correlation of geometry and calculation (the retrospective explanation of the algorithm for calculation in terms of a geometrical argument). Both are typical to what we referred to earlier as "deuteronomic" practices – the rearrangement of previously known facts, to supply them with structure and context. That Al-Khwarizmi's project is primarily deuteronomic can, in fact, be shown. His approach is essentially based on stating problems in the terms of given numerical terms. Thus, it is possible to compare such numerical terms to other problems, preserved in earlier cultural traditions. It is then found that Al-Khwarizmi's problems were widespread in the Ancient Near East, from Babylonian times onwards. Hoyrup – who made the fundamental studies concerning the history of such problems – suggests the following account of their nature.[11] Practitioners of calculation would enjoy exercising their skills in the more ludic form of *puzzles*. Thus they would approach each other with questions such as, say: "The square field with ten times its side is thirty-nine. How much is the field?" Occasionally such puzzles get written down, perhaps because they enter some education curriculum, or just because someone wishes to compile a list of such puzzles; but their natural form, as well as transmission, is oral. (The practitioners themselves, however, might well be literate: this is comparable to the way in which jokes, and indeed puzzles, circulate in our own highly literate society as basically oral texts.) In the Greek cultural context, such puzzles were of course still on the lips of practitioners, but they belonged to a different cultural stratum from that of elite Greek mathematics. The intellectual, aristocratic authors of Greek mathematics were different from the practitioners of calculation, and the demands of the Greek mathematical genre were very different from those of orally transmitted puzzles. Thus little contact was made between the two traditions, although the texts of Hero and of Diophantus, for instance, clearly betray an acquaintance with the tradition of calculation puzzles. Hero, in particular, seems to employ even a language reminiscent of such calculation puzzles: we recall his treatment of geometrical relations in terms of multiplication, and Archimedes' deliberate exploitation of this tradition. We can say, then, the following. In the Ancient Greek world, the two forms – literate geometry, and oral calculation puzzles – subsisted separately, with occasional contacts, sometimes exploited for deliberate effect, as we have seen in the case of Archimedes' solution to the

[11] Hoyrup (1996), and references there.

problem of areas and lines. In Eutocius, we saw how his project, to put together the various strands in the tradition bequeathed to him, within a single context, resulted in a certain blurring of such dividing lines. Eutocius set out to justify Archimedes' statements using the "area on line" expression, in terms of canonical Euclidean proportion theory. As a consequence, he ended up producing a piece of proportion theory that treated geometrical objects as terms in multiplication. Three centuries later than Eutocius, but working in a similar intellectual environment (there was very little taking place, in mathematics, between the two!), Al-Khwarizmi set out to accomplish a similar task – to bring together the various strands of the tradition he had available to him – and achieved a similar result. By providing geometrical "reasons" for the solutions of calculation puzzles, Al-Khwarizmi had, explicitly, to treat geometrical relations in terms of multiplication.[12]

Does Al-Khwarizmi already achieve something like a theory of equations? Our goal here is not to go into a terminological debate concerning the meaning of "equation," or into the equally empty game of assigning priorities. But it is important to note how Al-Khwarizmi does indeed move towards equations – as well as how he does not.

It would be obvious that Al-Khwarizmi's route is, in a sense, the opposite to that of Eutocius. Eutocius starts from geometry, and finds himself reaching towards calculation; Al-Khwarizmi starts from calculation, and finds himself reaching towards geometry. Indeed, we should point out the similarity between the history of calculation and the history of geometry. We have noted the "aura" characteristic of ancient geometrical problems: they were not designed to serve as parts of a system, but as unique results of intrinsic value. The same, of course, is true of calculation puzzles. A puzzle is unique. When it comes to be seen as a special case of a generally applicable mechanism, it loses its ludic aspect, since there is no longer a quest involved in the finding of the solution,

[12] We may ignore here the question of which cultural strands Al-Khwarizmi had available to him, precisely. He may have been drawing on any combination of Hindu, Babylonian, Egyptian, and Greek science, transmitted possibly via Syriac, Persian, or other languages. See e.g., Youschkevitch (1976) 34–51, Gandz (1936) for various suggestions, with references.

merely a mechanical application of a rule. From *puzzle*, it is transformed into an *exercise* (which is the form in which algebraic problems are treated in modern schools – and, apparently, were treated in some Babylonian scribal schools!). This makes it immediately obvious why ancient calculation puzzles are characterized by special numerical terms, and are not treated as particular cases of general formulae. Whether or not the ancient practitioner was capable of finding general formulae, this would have been alien to the spirit of the game. Al-Khwarizmi, however, intrudes upon this ludic world with a subtly different attitude. His aim is not to encounter problems one by one, or even as a collection of separately intriguing problems, but, instead, to systematize them: he wishes to find a classification under which all problems of a given type can be subsumed. This very interest in systematization changes the nature of the mathematical object studied: instead of the mathematical object, e.g., the problem "The square field with ten times its side is thirty-nine. How much is the field?" one can have a different kind of mathematical object, e.g., the equation "Quantities and roots equal to numbers."

Does Al-Khwarizmi offer, then, a theory of equations? Not yet: while he adumbrates the concept of the equation, all his problems are raised and solved *as problems*. It is likely, after all, that Al-Khwarizmi did not produce, in the detail of his mathematical exposition, much original contribution. The problems, as well as their solutions, were mostly given by his sources. Like Eutocius' commentary, Al-Khwarizmi's algebra was, ultimately, a fairly unambitious work. Once again, however, we come across the irony of limited ambition, translated into major transformations. Without himself doing anything beyond classifying the results of the past, Al-Khwarizmi, effectively, created the equation. Starting from his work, it would be possible for later mathematicians to treat problems as equations.

This may have happened more than once. Sesiano discusses a manuscript dated to 1004/5 where, following a discussion of the problems solved by Al-Khwarizmi, the (anonymous) author moves on to note:[13]

[13] Following Sesiano (2002) 200.

> Case of compound equations involving three elements not in continued proportion, or more, either in continued proportion or not. This is the case for the two possible categories involving three elements . . . These do not admit of a treatment with our above algebraic resolutions [i.e. by Al-Khwarizmi's methods], but only a geometrical one using conic sections . . . [and the author moves on to a brief list of such cases].

The system of soluble problems made canonical by Al-Khwarizmi has, as it were, the germs of its extension into a wider study, where equations and their conditions of solubility are studied as such. The crux of this discussion would have to be the point where Al-Khwarizmi's methods face resistance: the third degree or, in other words, problems akin to Archimedes'.

Al-Mahani, in particular, suggested conceiving of Archimedes' problem as an equation. As we have seen in the previous subsection, he was unsuccessful in his attempted solution. But other Arabic mathematicians were successful, and Khayyam, finally, would incorporate his own solution into his own algebra – his much more ambitious attempt to provide what we may now call, with all justification, a theory of equations. Let us then move on to consider the final transformation of Archimedes' problem in the work of Khayyam.

3.3 Khayyam's solution within Khayyam's algebra

In an early treatise by Khayyam, *The Division of a Quadrant of a Circle*, the author sets out (fig. 14) to find a point G on the quadrant AB so that (with GH perpendicular to EB) AE:GH::EH:HB. Geometrical analysis reduces this problem to a set of relations holding within right-angled triangles which, by assuming (for the sake of the argument) certain arbitrary values, transforms into the following relation: *A cube and two hundred things are equal to twenty squares and two thousand in number.* Having stated this, Khayyam proceeds to a long excursus on the state of the art of algebra, making many explicit bibliographic references. He points out the need for an exhaustive algebra, one that will deal with all problems involving terms up to the cube (but ignoring more complex, fictional terms such as *square-square*, etc.).[14] We see

[14] *The Division of a Quadrant of a Circle* is edited in Rashed (1999).

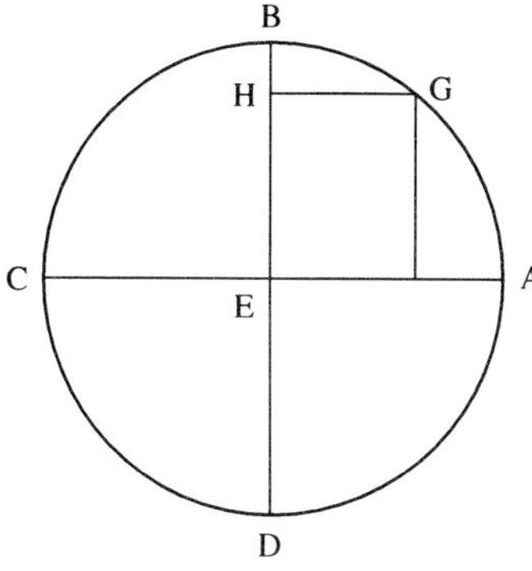

Figure 14

therefore that in the three centuries separating Khayyam and Al-Khwarizmi certain developments had occurred. Al-Khwarizmi's treatment reached only as far as a single multiplication between terms (his "quantity," which was the result of a multiplication of "roots"). This was extended by later algebraists to any number of multiplications, which were then systematically correlated with geometrical terms: the cube, and then (without obvious geometrical sense) the square-square etc. Khayyam wishes, on the one hand, to take the correlation seriously, so that the very meaning of algebraic terms would depend on the existence of geometrical correlates and, on the other hand, to make the treatment of algebraic problems as far as the cube systematic.

In other words, while much has happened between Al-Khwarizmi and Khayyam, much remained the same: the urge to correlate; the urge to systematize. In *The Division of a Quadrant of a Circle*, the correlation and the systematization were merely a programmatic statement. The fulfillment of the program is in Khayyam's, later, *Algebra*.

In what follows I offer a number of observations on this treatise by Khayyam.

A basic feature of the treatise is the central role played in it by introductory statements. Reflections upon the treatise, and the treatise itself, form a continuous whole. Khayyam's *Algebra* is marked by a strong, explicit setting in a historical, bibliographic, philosophical, indeed even an autobiographical context. This setting is not a marginal, "coloring" addition to the work, but a fundamental constituent, and indeed "setting" and "work" are hard to tell apart.

Let us look at the introduction, then.[15] Khayyam begins his treatise by putting his subject-matter in its philosophical, *systematic* position. "One of the scientific principles required in that part of wisdom known as 'mathematical' is the art of Al-Jabr wa l-Mukabala . . . and in it, there are kinds in which one requires kinds of preliminary propositions which are very hard, and whose solution inaccessible to most researchers."[16] So: (A) wisdom, within it (B) "mathematical," within it (C) "Al-Jabr wa l-Mukabala," within it (D) "kinds" which are especially difficult: it is this, fourth layer of systematic analysis to which Khayyam's treatise is dedicated. As can be seen, the systematic position immediately gives rise to a mathematical, or even bibliographic position (the kinds "require preliminary propositions") as well as a historical position (the solution was "inaccessible to most researchers"). It is to this historical context that Khayyam now proceeds, noting first the absence of ancient (i.e. Greek) extant works, then the limited success of later (i.e. Arabic) works. This historical notice is of special interest, as Khayyam mentions explicitly the Archimedean problem: Al-Mahani tried to solve it without success, Al-Khazin then solved it. This is about as much as the moderns achieved, according to Khayyam, until his own time. Thus the historical context leads smoothly to the autobiographical context: Khayyam tells us about his lifelong desire to study this field, the obstacles put in his way – not least by some obnoxious people. Finally he tells us of his studies with Abu Tahir (for whom he has very warm words) and of his ultimate success in producing this work: the historical route, from Archimedes, through Al-Mahani, Al-Khazin and Abu Tahir, ends with Khayyam himself.[17]

[15] For the following analysis of the introduction, compare Rashed (1999) 117–25.

[16] All the translations offered here from Rashed's *Algebra* are based on a combination of Winter and Arafat (1950), with Rashed (1999) (which I take to be the authoritative version of Rashed's translation). I sometimes deviate from both, mainly to accommodate the text to my terminology used in the translations from the Greek. Obviously, readers should assume that, whenever my translation conflicts with Rashed's, his is the more precise rendering of the Arabic.

[17] As will be noted below, introductory material keeps being provided later in the work, including historical context: this is done in particular in an excursus added at the end of

With this personal note, it would seem that the introductory material was over; but Khayyam presses on with a more detailed mathematical-philosophical positioning of the field. Al-Jabr wa l-Mukabala is defined; the quantities it deals with are enumerated and analyzed, from both metaphysical and mathematical points of view (typically, two previous authors are mentioned in this context: Aristotle and Euclid). Khayyam then discusses further the scope of the specific field he deals with: as we have been led to expect from the very start, this is done by reference to the preliminary propositions required, i.e. here arrives the bibliographic context. This is a set of three works: Euclid's *Elements* and *Data*, as well as Apollonius' *Conics*. Readers are warned not to attempt the treatise without a previous mastery of this background.[18]

Khayyam's introduction does not stop there, and now he goes on to discuss the nature of algebraical equations, from metaphysical and mathematical points of view, and this survey leads on, very naturally, to a survey of the types of equations studied in this field. This survey, finally, constitutes what may be considered the treatise proper. The language gradually becomes now that of Greek-Arabic geometry and algebra, with figures lettered by the Arabic alphabet, and the language of theorems and proofs. Notice, however, that the early types of equations dealt with are very simple, they do not call for detailed mathematical discussion. Thus the continuity between "introduction" and "treatise" is further stressed: the text, even in its more mathematical part, starts out as relatively "discursive," ordinary scientific Arabic, and only gradually it becomes more specifically mathematical. Finally, even the later part of the work – which contains many complex mathematical propositions, naturally in the mathematical mode of exposition – more general, discursive remarks are frequently made. For example, Khayyam

the work, Rashed (1999) 227 ff. Further information, particularly on Ibn Al-Haitham, is mentioned towards the end of the work proper, Rashed (1999) 223–5; while many other references to "previous," unnamed mathematicians are made throughout the work, e.g., Rashed (1999) 197. Finally, in an interesting complication, Khayyam refers to a treatise by himself, in Rashed (1999) 129.

[18] The bibliographic coordinates of the work keep being provided later on in the treatise: quite frequently, Khayyam refers explicitly to propositions from the three books mentioned, naming book and proposition as the authority for a certain claim: I count at least nineteen such references in the work.

systematically describes the type of equations dealt with, in more general terms (e.g., whether or not it has cases). Also, when such comments suggest themselves, he notes the relation between his works and earlier works. We shall see discursive remarks of this kind in the text quoted in the following subsection; though it should be noted that this text is one of the least discursive ones in the treatise.

Briefly, then, Khayyam's treatise is characterized by a seamless transition from general, contextual comments, to the mathematical results themselves. Indeed, the context – setting out the results as belonging to a certain system – is not some marginal comment, but is the key to the work, which is all about setting out cases. Thus the introduction, in a real sense, never ends. It is typical that the word "introduction" is supplied by the modern editor:[19] it is not in the original, because the original is not neatly divided between "introduction" and "text." The work, as it were, is not just algebra, but also "An introduction to algebra."

The central role of the introductory material is related, as we see, to another important feature of this treatise, namely its strong interest in systems of all kind. Khayyam is constantly interested in articulating domains: dividing them, and organizing them according to some overarching principles. This indeed is the very start of the work, with its species-genus arrangement: (A) wisdom, within it (B) "mathematical," within it (C) "Al-Jabr wa l-Mukabala," within it (D) "kinds" (of a more difficult nature). This Porphyry's tree is but the first of many lists and divisions made in the treatise. In history, people are either "ancients" or "later."[20] In the metaphysics of algebra, its objects are "the line, the plane, the solid, and time"[21] – tellingly, Khayyam immediately refers to Aristotle's *Categories* (as well as to a *Categories*-based comment in the *Physics*). Khayyam lists the "degrees":[22] thing, square, cube, square-square, etc.; towards the end of the treatise, he reverts to the same list, now to produce it together with its correlate, list of "parts" ("part of a square" is what we would call "1 over square": if the square is 4, part of the square is $\frac{1}{4}$). The one-dimensional list of degrees

[19] Rashed (1999) 117.4.
[20] Rashed (1999) 117.
[21] Rashed (1999) 121.
[22] Rashed (1999) 121.

thus becomes a two-dimensional grid and, in acknowledgment of that, Khayyam explains that he decided, for clarity's sake, to set out the "parts," together with the original degrees, in a table.[23] Now, similar tables form what may be considered the heart of the treatise. Near the beginning of the work, following other divisions concerning algebra, Khayyam sets out the various kinds of equalities.[24] These form, once again, a many tiered genus-species structure: equalities are either "simple" (binomials) or "complex" (polynomial). "Complex" equalities are either with three, or with four terms (note that Khayyam does not deal with degrees beyond the cube: this results from his deeply geometrical conception of mathematics, to which we shall return in the next section). For several of the species obtained in this manner, Khayyam distinguishes further species (e.g., between equations that were treated by earlier mathematicians, and those that were not), so that finally each *infima species* contains no more than a few equalities (six at most). The bulk of the treatise is an unpacking of this preliminary list: a set of solutions of those equalities, always following this genus-species structure. Overarching division is thus, quite simply, what the book is about. Tellingly, even the *names* of the species and genera derive from the list, as they are called, e.g., "<the kind of> the six kinds" and so on.

There are many further divisions and lists made throughout the book, in the course of the mathematical argument itself. Several distinctions occupy Khayyam explicitly. Most important is the distinction between problems that do not require conic sections from those that do. This, indeed, is the main division of the book:[25] following a list of problems and solutions which do not require conic sections, Khayyam makes a break in the argument. "After introducing these kinds that could be proved from the properties of the circle, that is from the book of Euclid, let us discuss now the kinds that cannot be proved except with the properties of the <conic> sections." (Note, incidentally, how mathematical and bibliographical distinctions coincide.) The break is very noticeable in the overall structure of the book as, for once, Khayyam deviates from the structure set out by the division of equalities, and introduces

[23] Rashed (1999) 219. [24] Rashed (1999) 125–9. [25] Rashed (1999) 153.

further auxiliary lemmas on solid figures.[26] Another crucial distinction for Khayyam is that between problems that are always soluble, and those that are not: those distinctions do not divide the book neatly, as the circle/conic sections division does. Thus, Khayyam makes those distinctions case by case: following each kind of problem, he notes whether or not they are always soluble. Thus, e.g., at the end of "the fifth kind of the remaining six kinds of three terms" Khayyam notes that "this kind has different cases, some of which may be impossible," while at the end of the next kind he notes that "this kind has no different cases, and none of its problems is impossible."[27] In other words, the treatise sets out to impose three separate grids on the universe of algebraic problems: the grid defined by number of terms and their relations (the one set out at the original table); the grid defined by the mathematical/bibliographic distinction of circle from conic sections; and the grid defined by the presence or absence of impossible cases. We see that one of the explicit interests of Khayyam is to investigate the pattern of this triple superposition.

The impulse to divide and to list goes, however, well beyond those basic grids. The work is articulated, throughout, by comparisons and parallel parts. In some simple cases, having offered one proof, Khayyam often moves on to offer another one, alternative to it. Once again, this is often explicitly marked according to a preconceived grid. The sixth kind, for instance, is defined, and then immediately we have the words "proof by numbers," followed by a very brief proof; and then "by geometry," and another brief proof follows.[28] In some other, more complex cases, the nature of the problem makes it natural to distinguish, not kinds of proofs, but kinds of situations arising within a single proof. In the next subsection, for instance, we shall see a case where Khayyam distinguishes three possible configurations that may arise from a single geometrical situation. Typically, the distinction is made explicit, and is even marked out in the layout of the work, as the three figures are labeled "first," "second," and "third."

Thus different proofs, and different cases within proofs, are put side by side. Further, Khayyam puts side by side proofs and

[26] Rashed (1999) 155–61. [27] Rashed (1999) 181–3. [28] Rashed (1999) 135.

examples. Consider again the table setting out "degrees" and their corresponding "parts." We may now notice that, besides the named degrees and parts, the table also lists numbers: those numbers are examples of such degrees and parts (taking 2 as the basic root). Such articulations of general proof, and of particular numerical examples, are often repeated through the work. In some cases, Khayyam uses a particular example instead of a general proof. For example, instead of solving generally the case of "a square equals a number of roots," Khayyam simply offers a special case, "a square equals five times its root," allowing the general solution (the root is equal to the number of roots mentioned in the problem) to be apparent from the particular case.[29] In some other problems, general statement exists alongside a particular example, as in the immediately following problem, "<a number of> things equal a cube." Khayyam explains explicitly that this general problem is essentially like the problem "a number equals a square." This is explained as follows: "example: four roots are equal to a cube; it is like has been said: four, a number, is equal to a square."[30]

Notice how, in the text quoted above, the word "example" is used explicitly – a sort of local signpost. The articulation of the work is never implicit. Indeed, as the same example also shows, the structural features of the treatise – how its different parts relate to each other – is always an interest of Khayyam. Thus different problems are related, in what may be considered, anachronistically, a "reduction": one problem is shown to be equivalent to another. Thus Khayyam states explicitly that a certain species of problems is all equivalent to another, and then proves this equivalence, each time using particular examples, sometimes to substitute the general argument, sometimes to corroborate it.[31]

The word "example" is one kind of local signpost used to articulate the work; other words are used as well, such as "by numbers" and "by geometry" which we have seen already, as well as, simply, "proof": that is, here and there, following a general statement, Khayyam would introduce his mathematical argument by the single word "proof." (By my counting, this minimal title occurs ten times in the treatise, though I may have missed some occurrences.)

[29] Rashed (1999) 133. [30] Rashed (1999) 135. [31] Rashed (1999) 147–53.

This is correlated with several expressions similar in meaning to QED: "and that's the goal," "and that's what we wanted to prove," etc.; I count twenty-four occurrences of this expression.

By far the most important signpost is, of course, the word "kind," followed by an ordinal, and often introduced by a connector. So, for instance, a problem is introduced by the words "And the second kind of this."[32] This constant repetition of the word "kind" is the main structural feature of the work, and may well have been so even at the visual level. While of course no autograph survives of the work, at least one manuscript (BN Arabe 2458) systematically sets out the expressions containing the word "kind" in bigger characters: this has a marked visual impact.[33] (Note that this kind of visual articulation is common in many Arabic scientific manuscripts, though sometimes using color instead of size.) Finally, in some parts of the work, a similar effect of articulation is obtained by the figures, which are (as is the standard elsewhere) positioned near or at the end of their respective problems, thus enhancing visually the verbal articulation of the work.

So the work is characterized throughout by an impulse to divide, to articulate, to put into systematic structure. To complete our observations, a final feature of the treatise must be added: the impulse is, often, not merely to articulate domains, but fully to exhaust such domains. Once again, it is instructive to take a non-mathematical example, namely the historical excursus. In surveying the domain of previous works in his field, Khayyam proceeds by an exhaustive division into "ancients" and "moderns," and then reasons as follows for the ancients:

> We have no treatises from them concerning it: perhaps, after having studied and looking for it; they failed to grasp it; or their theories did not lead them to study it; or their treatises were not translated into our language.[34]

[32] Rashed (1999) 141. Notice how the system works as a whole: the word "kind" is the local signpost, signaling the start of a new problem; the word "and" positions the kind inside a sequential system; the word "second" provides the place in the sequential system; the word "of this" hints at the system being referred to. The whole expression, finally, is an unpacking of an entry from the original set of tables.

[33] This manuscript apparently is, according to Rashed (1999) 109–13, the copy closest to the autograph, though of course this does not guarantee that this particular feature is authorial.

[34] Rashed (1999) 117.

What we see here is Khayyam's urge to obtain truth by encompassing a domain of possibilities. This immediately becomes a defining feature of the work. The subject-matter, algebra, is exhaustively defined, in many aspects. The kinds of quantities, as we have noted, are enumerated, in an exhaustive list which – purely for exhaustion's sake – includes time in addition to the other mathematical quantities. (It is in this context that reference is made to the *Categories*, a work that Khayyam must have understood as an exercise in exhaustive systematization.) Then the various degrees are spelled out, from the root upwards (and, much later in the work, from the root *downward*, dealing with "parts"). Then Khayyam stops short the infinite expansion of degrees (to square-square, square-cube, cube-cube and beyond) by insisting on the geometrical meaningfulness of quantities: "since there is no other dimension [beyond cubes], the square-square and what comes beyond it are not among the magnitudes." Thus the exhaustive list of kinds of magnitudes helps to delimit an exhaustive set of kinds of degrees (number, root, square, cube), and this immediately leads on to the heart of the treatise, which is the exhaustive list of kinds of equations defined by those four degrees.

Thus at the most global level the treatise operates through exhaustive listing. But the same principle is operative in many individual proofs. This is the essence of Khayyam's interest in "cases" in proofs, which derive from some exhaustive list of a set of possibilities: "And these two <conic> sections will either meet or not meet."[35] Having made such an assertion, Khayyam then moves on to study each of the possibilities. Many proofs of the treatise are structured by such exhaustive lists, and we shall see an example in the following subsection. In an interesting complication, this example has a two-tiered exhaustive classification (within a certain possibility, further sub-possibilities are surveyed). Exhaustive lists, that is, can become complex, many dimensional systems.

It should be noted that this interest in argument through exhaustive lists is remarkable, given the subject-matter taken by Khayyam. In the terms of Greek mathematics, Khayyam deals almost exclusively with problems: that is, he defines situations, and sets himself

[35] Rashed (1999) 167.

the task of finding lines satisfying the definitions. Now, argument through exhaustive lists is often used in Greek mathematics – but mainly in two contexts. One is that of *reductio* arguments, which work through the exhaustive principle that P or not-P, showing that P is impossible and thus deriving not-P. Another – essentially a development of *reductio* arguments – is what is called (for other reasons) "the method of exhaustion." There, it is argued that a certain object is either greater, smaller, or equal to another one; the "greater" and "smaller" options are ruled out and the "equal" is thus proved. Both *reductio* arguments, and the "method of exhaustion", are useful, for obvious reasons, not for *problems*, which achieve a task, but for *theorems*, which state a truth. Finally, a very special work within the corpus of ancient Greek mathematics (but one in which Arab commentators had a special interest) does work through the principle of classification: this is *Elements* Book x.[36] This book classifies the kinds of relations of incommensurability. Once again, however, classification is used in the context of theorems. (Furthermore, the classificatory object of the work remains mostly implicit.) It is a peculiarity of Khayyam's argumentative style, then, to rely so heavily on exhaustive lists in a treatise dedicated to *problems*.

But then again, exhaustive lists is what this treatise is about: Khayyam's main claim is not that he proved this or that result, solved this or that problem, but that he encompassed an entire domain. The goal of the treatise is totality: thus to claim that an object has a certain position in the system is not some tool used for listing objects, a mere signpost. The signaling of positions in a system is a tool used in the exhaustive survey of the *entire* system. Each separate part of the treatise – each case within a problem, each kind of equality, each group of kinds – participates simultaneously at two levels. At one level, it makes a specific claim, separate to it; at another level, it functions in an overall argument, surveying the domain of algebra.

To sum up, then, we saw three structural features of Khayyam's *Algebra*. The first was an inter-penetration of the introduction, and the treatise proper: the treatise was a direct continuation of the

[36] For its reliance on the principle of classification, see Vitrac (1998) 51–63.

introduction, since the treatise was simultaneously, *in* algebra, and *about* algebra. The second was the strongly articulate, systematic nature of the treatise: it constantly arranged itself in various divisions and lists. Finally, we saw how the two features are connected through the principle of exhaustive lists. The interest of the treatise is in arranging claims – and objects – into systematic orders, so those separate claims become, simultaneously, components in a large-scale claim about the entire domain of algebra.

Having made those general observations on the treatise, it is time to see its part devoted to the Archimedean problem.

3.4 The problem solved by Khayyam

I offer here a translation, based on Rashed's and Vahabzadeh's important new edition and translation of Khayyam's work, of the problem in Khayyam's algebra which is the descendant of the Archimedean problem. It is no longer represented as such (even though, as his bibliographic references show, Khayyam was very much aware of the connection). According to its title, this is simply the fifth problem in a group of six problems of three terms; modern editors sometimes number the problems in this treatise, and then it becomes "Problem 17." There are altogether twenty-five problems, so this problem occupies an advanced position in the book.

The translation has no claims for style or precision. It is brought here so that we can discuss the text and, to make the comparison with Archimedes easier to follow, I adopt the same conventions adopted in my translation of Archimedes.

The fifth kind of the a "six remaining three-termed kinds": A cube and a number equal a square.[37]

(a) We suppose AC as the quantity of the squares; (b) we construct a cube equal to the given number, and let its side be H.[38] (1) And the side H will have to be either equal to the line AC, or greater, than it, or smaller. (2) So, if it is

[37] Sc. *a certain quantity* of squares.

[38] This point is rather confusing: the problem sets out a cube that, together with a number, equals a (multiple of a) square – the cube and the square being related in that they share the same side. Now, Khayyam immediately moves on to construct a further, auxiliary cube – not to be confused with the one set out by the problem itself – which is equal to the given *number*. Its side is H so that one may say that the given number equals (in modern symbolism) H^3.

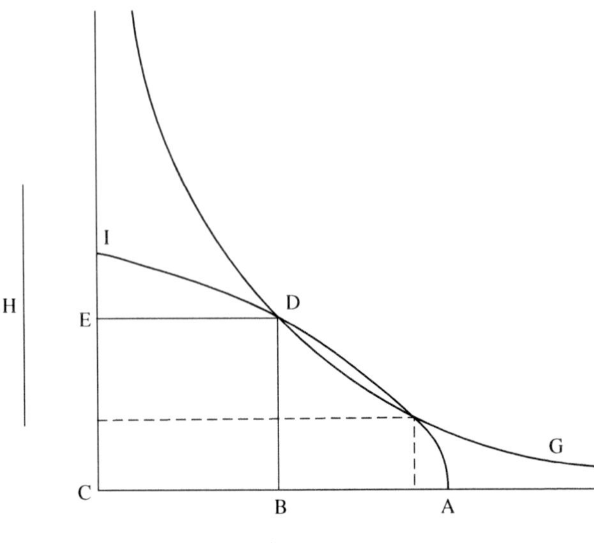

Figure 15

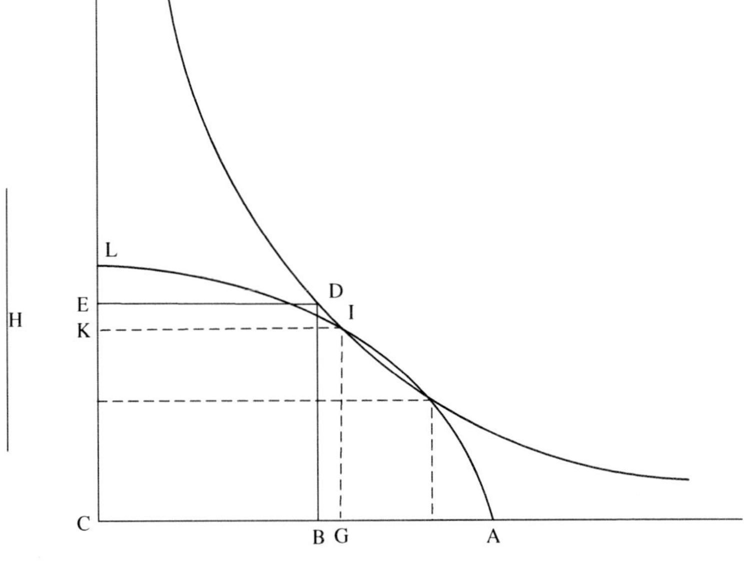

Figure 16

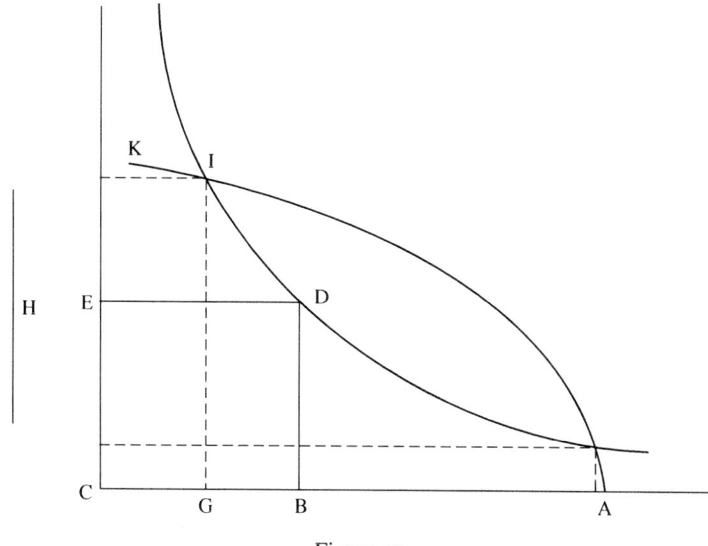

Figure 17

equal to it, the problem is impossible, (3) since the side of the required cube will have to be equal to H, or smaller, or greater. (4) So, if it is its equal, the product of AC by its <=the required cube's side> square is equal to the cube of H; (5) and the number shall be equal to a quantity of squares, and there will be no need to add the cube.[39] (6) And if the required side is smaller than it <=than H>, the product of AC by its <=the required cube's side> square is smaller than the given number, (7) so the quantity of squares will be smaller than the given number, even without the addition.[40] (8) And if the side is greater than H, its cube is greater than the product of AC by its <=the required cube's side> square, even without the addition, to it, of the number.[41] (9) Then, if H is bigger than AC, the impossibility

[39] I.e. the original equality is "cube with number equals quantity of squares," but we have "number equals number of squares," i.e. in effect, no cube – so obviously the problem is impossible (in our terms, it may be said that Khayyam does not consider zero to be a solution to the problem).

[40] I.e. the original equality is "cube with numbers equals quantity of squares," but we already have "number greater than quantity of squares," and adding in a cube to the number will not make it any smaller! (In our terms, it may be said that Khayyam does not consider negative numbers as solutions to the problem.)

[41] I.e. the original equality is "cube with number equals quantity of squares," but we already have "cube is greater than quantity of squares," and the addition of a number can only make this worse. (In our terms, it may be said that Khayyam does not consider negative numbers as possible parameters.)

Steps 3–8 are all governed by Step 2, and together show the impossibility of the case H = AC.

in the three cases shall be even greater.[42] (10) So it shall be necessary that H will be smaller than AC, and otherwise the problem is impossible.

(b) So we cut BC, equal to H, from AC. (11) So the line BC shall be either equal to AB, or bigger than it, or smaller. (c) So let it be, in the first diagram, equal to it; (d) and in the second, bigger than it; (e) and in the third, smaller than it. (f) And a square DC shall be completed in the three diagrams, (g) and we produce, at the point D, a hyperbola, asymptotic to AC, CE, (h) which is DG in the first diagram, (i) DI in the second and the third. (j) And we produce a parabola, whose vertex is the point A, and whose axis is AC, and its *orthia* is BC;[43] (k) which <parabola> is AI in the first diagram, (l) and AL in the second, (m) and AK in the third. (12) And the sections shall be known in position.[44] (12) So in the first <diagram>, the parabola passes at the point D, (13) since the square of DB is equal to the product of AB by BC:[45] (14) so D shall be on the perimeter of the parabola; (15) and it <=the parabola> will meet <the hyperbola> at another point – which you can grasp, with the least thought.[46] (16) And in the second, the point D shall be outside the perimeter of the parabola, (17) since the square of DB is bigger than the product of AB by BC.[47] (18) So if the two sections meet, by a tangency at another point or by an

[42] Khayyam intends that we verify by going through the previous three cases, which the reader may now do. This Step 9 shows the impossibility of the case H>AC so, together with Steps 2–8, the ensemble of Steps 2–9 shows that the only case which may at all be possible is H<AC, as asserted in the following step.

[43] *Orthia* is the formulaic Greek expression, literally meaning something like "the rightish <line>," transformed in Arabic into the equally formulaic expression "the right side," and which I finally transliterate back into the original Greek, to suggest the formulaic ring of the expression in Arabic. We have seen this term in chapter 1 above: it refers to the line, defining a parabola so that – applying modern terms to, e.g., diagram 1 of this proposition – every perpendicular from the parabola on the axis, such as DB, satisfies $DB^2 = (orthia)*(BA)$ or – as this construction stipulates – $DB^2 = (BC)*(BA)$.

[44] The claim is that a hyperbola is determined by a point through which it passes, together with its two asymptotes (*Conics* II.4), while a parabola is determined by its vertex, axis, and *orthia* (*Conics* I.52).

[45] The square of DB is the square EDBC and, by the definition of diagram 1, AB = BC and so $AB*BC = BD^2$; by a converse of *Conics* I.11, the parabola must therefore pass at the point D.

[46] Once again, Khayyam addresses the reader with an "exercise," this time curiously explicit. The truth of the claim is visually compelling, but good ancient and medieval authors would prefer not to rely on the diagram for exploring the relations of conic sections, as these were drawn (intentionally) falsely, by arcs of circles (more of this in section 3.6 below). I have given the matter a little thought, and then some more thought, and finally I think as follows: if the two sections cut each other at D, the claim is indeed obvious (for the hyperbola will have to "escape" from inside the parabola, so as to avoid cutting the asymptote). The two sections cannot be tangent at D, since this would imply that, with the tangent produced, it should be cut into equal segments at the touching point (*Conics* II.3), which in turn would imply that CB, that is DB, is equal to the segment from B to the cutting-point of the tangent and of the line BA produced; but DB is already equal to BA and an impossibility arises.

[47] And it is at the point on the line DB, where the square is equal to the product, that the parabola passes: a converse of *Conics* I.13.

intersection, then the perpendicular drawn from this <point of meeting> will have to fall between the points A, B; (19) and the problem is possible; (20) otherwise it is impossible.

This tangency or intersection was not grasped by Abu'l-Jud, the eminent geometer, so that he reached the conclusion that if BC is bigger than AB, the problem would be impossible; and he was wrong in this claim.

And this kind is the one that baffled Al-Mahani (among the six kinds).[48] So that you shall know.

(21) And in the third diagram, the point D shall be interior to the parabola,[49] so the sections cut each other at two points.[50]

(n) And, in all, we draw, from the point of meeting, a perpendicular on AB, (o) and let it be, in the second diagram, IG; (p) similarly, <we draw> from it <=D> another perpendicular, on CE, namely IK. (22) So the rectangle IC is equal to the rectangle DC,[51] (23) so the ratio of GC to BC shall be as the ratio of BC to IG.[52] (24) And IG is among the ordinate lines in the section AIL;[53] (25) so its <=IG> square shall be equal to the product of AG by BC. (26) So the ratio of BC to IG is equal to the ratio of IG to GA.[54] (27) So the four lines are proportional: the ratio of GC to CB as the ratio of CB to IG, and as the ratio of IG to GA. (28) So the ratio of the square of GC, the first, to the square of BC, the second, as the ratio of BC, the second, to GA, the fourth. (29) So the cube of BC – which is equal to the given number – is equal to the solid whose base is the square of GC, and its height GA. (q) And we add the cube of GC as common; (30) so the cube of GC with the given number is equal to the solid whose base is the square of GC, and its height AC,[55] (31) which <=AC> is equal to the given quantity (of squares).[56] (32) And this is the goal.[57]

48 It is not altogether clear which "six kinds" are referred to: they could be either the six kinds to which this kind belong in Khayyam's treatise, or some six kinds Al-Mahani was baffled by. The reference to Al-Mahani, at any rate, is a follow-up to the brief mention in the introduction, where it was also mentioned that Al-Mahani studied the Archimedean problem; this is as much as Khayyam says explicitly to connect this problem with Archimedes.

49 The same reasoning as used in Step 17.

50 The hyperbola now needs to "escape" from inside the parabola, in *both* directions.

51 *Conics* II.3   52 *Elements* VI.16.

53 I.e. it is one of the lines defined in such a way that the square on them is equal to the rectangle contained by: (1) the line they cut from the axis, and (2) the *orthia* (*Conics* I.13).

54 *Elements* VI.16.

55 The cube of GC is, in fact, the solid whose base is the square of GC, and its height GC. Add it to the solid whose base is the square of GC, and its height GA, and you have a new solid, whose base is the square of GC, and its height (GC + GA). GC + GA is the same as AC, hence "the solid whose base is the square of GC, and its height AC," mentioned by Khayyam.

56 AC was set down as the quantity of squares, in the very first Step a.

57 We have produced a line – GC – whose property is that: its cube, together with a given number, equals a given number of its squares.

(33) Analogously with the two remaining cases, (34) except that the third has to give rise to two cubes, (35) since each perpendicular cuts from CA the side of the cube, (36) as has been proved.

So it has been proved that this case has different cases, some may include impossibilities, and it has been solved by the properties of two sections, both a parabola and a hyperbola.

3.5 Khayyam's equation and Archimedes' problem

We may now go back all the way and compare this treatment by Khayyam to the original solution we have seen by Archimedes, in section 1.2 above. Clearly, much changed. And yet, there was no break in history: no deep conceptual divide separating Archimedes from Khayyam. In fact, reading closely, one is at times struck by the degree of continuity between the two treatments, at times struck by Khayyam's originality. By delineating the lines of difference and similarity, then, we may obtain a finer understanding of the sense in which Khayyam's work was an "algebra."

In analyzing in chapter 1 the mathematical significance of the solutions offered by the Greek authors – Archimedes, Dionysodorus, and Diocles – I often tried to give an account of the possible route of discovery leading to the solution. The three authors were largely independent of each other, and so they had to discover their solutions on the basis of little background information. Thus one can make plausible guesses, on the basis of the solution offered itself, as to how it could have been found. The author faced raw geometrical reality and transformed it into a statement in words, and the words still have impressed on them this fresh stamp of reality. The same is no longer true for Khayyam. While we do not know exactly which works Khayyam was aware of, we know – from his own words – that he was acquainted with several treatments, some successful, some not, of the very same problem. Thus, when Khayyam sets out to produce his new version, he faces not geometrical reality in the raw, but geometry already transformed into verbal forms. He did not transform reality into words, but words into words, and since we do not know the precise words Khayyam had available to him in his tradition, we cannot work back this transformation to find his own way of reaching his formulation. But while we cannot enter Khayyam's mind in this way, of uncovering his mode of discovery, we can still concentrate on his mode of presentation: what are the stylistic features of Khayyam's writings, and what do they suggest for his conception of geometrical objects?

I start with a detail of Khayyam's exposition that is very typical of a certain duality (in a sense, continuous with Archimedes himself) – conjuring non-geometrical possibilities, while manifesting a sustained geometrical conception of the problem. I refer to Khayyam's system for naming lines.

To begin with, notice how my translation of Archimedes is peppered by the phrase "the <line> AB." This is a very minimal Greek expression, in transliteration: *hē AB*. The Greek definite article, in its singular feminine form, followed by two Greek letters (or, less frequently, three letters), for letters standing at points on the line. The Greek words *eutheia grammē*, "straight line," are dropped. (In my translation, I insert back the word "line," alone, inside pointed brackets.) However, these words are understood: the expression is merely a way of referring to specific lines in a specific figure. While my translation is no doubt irritating in its plethora of pointed brackets, those pointed brackets do serve a function in reminding us how much the Greek reader fills in, and how much it is felt that the text refers throughout to geometrical objects.

Khayyam's text is different, and this particular formulaic form is dropped altogether. This is indeed natural in a translated context: the easiest way to render the Greek *hē AB* in another language is simply by AB, if only because the expression *hē AB* contains nothing to translate besides definite article and Greek letters. Further, Arabic does not possess a declension of the definite article: but without the feature "feminine" spelled out on the definite article, it loses even the minimal meaning it had in the Greek. Finally, even phonologically, the expression "the AB" is problematic in Arabic, in which the definite article joins with the noun it governs to form a single word. The hypothetical expression *al-AB would be particularly strange, as the definite article would have to combine with a peculiar, extra-linguistic object – the letters of the diagram. Such linguistic speculations aside, it is clear that Khayyam's text differs from Archimedes' in its avoidance of this particular formula – with which go many other, more complex formulae. To put it simply, my translation of Khayyam contains far fewer pointed brackets than my translation of Archimedes.

This could have a consequence for the way in which geometric objects seem to be understood in Khayyam's text. Of course Khayyam occasionally does refer explicitly to lines as "lines" – as indeed Greek mathematicians also do. But Khayyam would drop this explicit reference in the contexts where a Greek mathematician would use only the abbreviated form *hē AB*. Thus the text would seem to speak not about lines as such but rather about objects represented by diagrammatic letters. In an expression such as, e.g., "(9) Then, if H is bigger than AC . . .", the "bigger" relationship holds, as far as the text is concerned, not between lines as such, but between such objects as are designated by diagrammatic letters. Since Khayyam does belong to a world where such letters can be used in calculation, and not only in geometry, his expressions now allow for a systematic ambiguity. Thus, for instance, the expression often used by Khayyam, "the square of AB," is truly indeterminate: it can refer both to the square (in terms of calculation) of the magnitude AB, or to the square (in the geometrical sense) produced from the line AB. It is indeed interesting to note that when Khayyam wishes to refer in non-ambiguous terms to a geometrical square, he does so by a different mode of naming of squares: "(f) and a square DC shall be completed in the three diagrams." By referring to the square through two opposite vertices, the reference can no longer be to "square" in the terms of calculation, and must be to "square" in terms of geometry. On the other hand, in some other expressions, the language of calculation seems dominant, as in, e.g., "(17) . . . the product of AB by BC." Archimedes would probably have "the <rectangle contained> by the <lines> AB, BC," but the absence of the formula "the <line> AB" makes it much more natural to refer to the product not as a geometrical, two dimensional object, but as a result of calculating with two symbolic objects, AB, BC.

And yet, while opening up these radically new ways of reading his text, it remains clear that Khayyam himself does not intend his text to be read in this way. There are many indications Khayyam conceives of his lines as geometrical configurations, and not as more generalized magnitudes represented symbolically.

The divide separating Khayyam from Archimedes is not conceptual.

Most simply, he operates upon his terms – even at the symbolic level of manipulation of characters inside the text – according to their geometric configuration. I mean the following. In keeping with Greek practices, Khayyam allows lines to be represented by the diagram, in whatever is the most natural way. Consider the line H: since it does not form part of the continuous geometrical configuration, it does not intersect with any other line and is thus not distinguished by any of its points. Thus it becomes natural to refer to it as a single unit (and not, as is done for other lines, through the points at its two limits). The result is that most, but not all lines in Khayyam's solution are two-lettered. This heterogeneous way of naming lines makes it somewhat less natural to see the expressions "H," "AC" as mere symbols. As mere symbols, they are homogenous; their heterogeneity is a function of the geometrical configuration.

The same grounding of the symbol in the diagram is seen in another phenomenon of Khayyam's treatise: the permutability of names. Again, as is also true of Greek mathematical practices, once a name is attached to an object it is generally kept the way it is. However, in a significant number of cases, names are allowed to switch: "(25) so its <=IG> square shall be equal to the product of AG by BC. (26) So the ratio of BC to IG is equal to the ratio of IG to GA. (27) So the four lines are proportional: the ratio of GC to CB as the ratio of CB to IG, and as the ratio of IG to GA. (28) So the ratio of the square of GC, the first, to the square of BC, the second, as the ratio of BC, the second, to GA, the fourth." In the course of these four steps – the key to the main geometrical property – AG has switched into GA, while BC has switched into CB and back again into BC. Thus the reference of those two-lettered objects cannot be purely symbolic – it is precisely their identity as symbols that such a permutation destroys. The identity of these objects is clearly given by the diagram where, indeed, it makes no difference whether you read them, as it were, from left to right or from right to left.

In short, we see that Khayyam opens up the possibility of considering his objects symbolically, as elements manipulated by the rules of calculation; yet essentially conceives of them as components in a geometrical configuration. This is seen at the most elementary level – the use of letters; but, as always, we encounter the same structural forces at all levels of analysis. For, after all, the entire treatise is determined by Khayyam's open-ended list of degrees – on into square-squares, square-cubes, and beyond; and his explicit decision, to limit himself to the four basic degrees alone. Most importantly, the same duality, with a preference to the geometrical, is shown in the kinds of mathematical statements and operations allowed. In this problem, we see Khayyam making a few claims whose geometrical significance is not apparent: "(a) We suppose AC as the quantity of the squares; (b) we construct a cube equal to the given number." What is the meaning of a line being "supposed as a quantity?" Or of a cube "being equal to a number?" Thus an equivalence between geometrical and more abstract objects is being suggested. However, those kinds of non-geometrical claims are limited to the stage of setting-out, where the general problem is set in geometrical terms. Following this setting-out, the argument proceeds strictly according to geometrical manipulations. None of the derivations made by Khayyam would have been inadmissible for Archimedes. True, Khayyam speaks of "product" where Archimedes speaks of "rectangle." However, Khayyam obtains his products through precisely the same geometrical techniques Archimedes could use for obtaining his rectangles: "(24) And IG is among the ordinate lines in the section AIL; (25) so its <=IG> square shall be equal to the product of AG by BC." Thus the difference between "product" and "rectangle" is in a sense no more than that of notation: in terms of admissible operations, Khayyam's terminology carries no consequences. Most tellingly, at the moment where Khayyam's treatment is most reminiscent of Al-Jabr wa l-Mukabala – when a quantity is added to two sides of an equation – there is nothing algebraic to his argument. "(29) So the cube of BC – which is equal to the given number – is equal to the solid whose base is the square of GC, and its height GA. (q) And we add the cube of GC as common; (30) so the cube of GC with the given number is equal to the solid whose base is the square of GC, and its height AC." The operation through which we obtain the equality

(solid whose base is the square of GC, and its height GA) +
(cube of GC) =
(solid whose base is the square of GC, and its height AC)

has nothing algebraic about it, and is instead Classical Greek cut-and-paste derivation, strongly based on unpacking information from the diagram. Apart from their strange initial formulation, then, Khayyam's proofs could be read, without perplexity, by any Greek mathematician.

But could they have been *written* by any Greek mathematician? While Khayyam uses the idiom of Greek mathematics, he also uses it in his own way, meaningfully different from, say, Archimedes'. At a mathematical, technical level, Khayyam's proof is clearly distinct from that of Archimedes. Let us try to analyze this sense of difference.

Once again, to have a sense of the difference, we should also notice the similarities. Both proofs, after all, are based upon an intersection of a parabola and a hyperbola, and both offer a study of cases, connecting it with the conditions of solubility. To some extent, such similarities may have historical explanations. Khayyam was well aware of at least some of the Arabic tradition, and at least some of it was aware of Eutocius' commentary: there is nothing unlikely, then, in Khayyam's solution being derived, ultimately, from Archimedes. There are also possible mathematical explanations for the similarity: the problem is after all the same; cubic equations are indeed equivalent to proportions involving lines and squares, and there are only so many curves that satisfy such proportions. We recall the similarity between the form of Archimedes' solution, and that of Dionysodorus; Ibn Al-Haytham's solution – whether or not dependent upon Archimedes – was even closer to the latter's treatment. In short: history and mathematics both determine a certain convergence between Archimedes and Khayyam.

This makes their divergence all the more apparent. This divergence has two aspects: the different roles played by the study of cases, and the different roles played, respectively, by ratios and equalities.

For the study of cases, consider Archimedes' discussion: "(2) To begin with, if it is greater,[58] the problem may not be constructed, as has been proved in the analysis; (3) and if it is equal, the point E produces the problem. (4) For, the solids being equal, (5) the bases are reciprocal to the heights, (6) and it is: as the <line> EA to the <line> AΓ, so the <area> Δ to the <square> on BE." For Archimedes, we see, the study of cases is simply a way of getting the main solution off the ground. In one case, the problem is insoluble, so this can be put aside, no further comment being made;[59] in another case, the solution is effected in a simple, direct way; so, having said that, the proof can unfold, without any further mention of cases being made.

Khayyam's solution is of course totally different. In mere quantitative terms, Khayyam's preliminary study of cases has 10 steps out of a total of 35 steps of the proof (29 percent), as against Archimedes' 6 out of 40 (15 percent). Indeed, the qualitative gap is wider, since Steps 3–6 in Archimedes' proof are not primarily a study of cases, but simply part of the solution: the division into cases serves not as an end, in this case, but as means for the solution. Thus we are left with Step 1 alone, which is a mere claim, not an argument, so that, in short, *Archimedes offers no argument whose end is the study of cases*. Khayyam, on the other hand, not only dedicates ten steps for this preliminary investigation: he goes on showing the same approach in the solution itself. We immediately notice that he offers not one, but three separate diagrams, corresponding to three possible geometrical configurations. And once again, these are not mere tools for obtaining the solution. Having made the necessary constructions and preliminary statements, Khayyam reveals the main interest of this study by division: "(18) So if the two sections meet, by a tangency at another point or by an intersection, then the perpendicular drawn from this <point of meeting> will have to fall between the points A, B; (19) and the problem is possible; (20)

[58] I.e. (<area> Δ, on the <line> AΓΔ) > (<square> on BE, on the <line> EA).

[59] Archimedes would then go on to a separate proof concerning the impossibility: we have read it in section 2.1 above. Notice, however, that (following the argument in section 2.2 above) Archimedes – in that proof as well – avoids any mention of cases. The proof simply unfolds for a single case. Instead of making his arguments through cases, Archimedes develops two separate, case-free lines of argument: one on the solution of the problem, the other on the conditions for solubility of the problem.

otherwise it is impossible." In other words, the configurations are simply another way of yielding cases of possibility and impossibility, so that the goal of this discussion as well – now well into the middle of the proof – is not the solution itself, but its division into cases. This is the heart of Khayyam's proposition – the moment where he stops to make historical and bibliographical statements, comparing his achievement with previous achievements. It is precisely such division into cases of which he prides himself: "This tangency or intersection was not grasped by Abu'l-Jud, the eminent geometer, so that he reached the conclusion that if BC is bigger than AB, the problem would be impossible; and he was wrong in this claim." Now, the next step in Khayyam's proof, 21, is another brief claim concerning cases; and then Steps 21–32 provide the geometrical argument concerning the solution, which is now seen as dependent upon the main claims. What Khayyam's solution at Steps 21–32 does, given its context, is not so much to solve a problem, but to show that *a solution is possible given a certain condition*. Finally Steps 33–5 wrap up the argument by suggesting how the same solubility may be seen for the other configurations.

Khayyam's proof, then, is not so much a solution to a problem, as a study of the cases arising out of the problem, arranged according to two exhaustive lists of equalities or inequalities:

(Content of Step 1): H $>/=/<$ AC
(Content of Step 11): BC $>/=/<$ AB.

The first part of the proof, Steps 1–10, studies the cases of possibility and impossibility arising from the first exhaustive list. The second part of the proof, Steps 11–35, studies the cases of possibility and impossibility arising from the second exhaustive list. The main geometrical property – Steps 21–32 – serves, in context, merely as an element inside this second study.

Khayyam looks at the problem, distinguishes its cases and studies them as items in an exhaustive list of equalities and inequalities; geometrical comments being made to the extent that they contribute to this study. Archimedes looks at the problem and develops its geometrical properties, realizing that these may also fall into different cases. This difference is one of the major reasons why Khayyam's problem feels more "algebraical" – why his lines tend to appear like sheer quantities. Since he plunges directly into cases and develops them before developing his geometrical study, he is bound to single out simple equalities or inequalities, which do not call for any geometrical imagination – the simple exhaustive lists of Steps I and II.

In his article "Steps towards the Idea of Function: a Comparison between Eastern and Western Science of the Middle Ages," Schramm commented on Khayyam's failure to study the point at which the parabola and the hyperbola are tangents. As we have seen, this point is exactly one third of the way above the given line – thus, an interesting property, which we would expect Khayyam to notice. Indeed, we have seen that Archimedes devoted his entire study of limits of solubility to this property. As Schramm put it:[60] "It is strange to find that 'Umar al-Kayyami does not mention this condition, already known to Archimedes. He likes to leave something for his readers to do." In fact Khayyam's silence on this point – as well as Archimedes' eloquence – are easy to explain. Since Khayyam's study of cases is logically prior to his study of geometrical properties, he is not interested in the geometrical properties of the points that define cases, as long as the points can be stated in terms of his exhaustive lists. For Archimedes, on the other hand, the cases are reached through an investigation of the geometrical properties of the configuration, hence he very naturally states the conditions for the tangencies of the sections. The different priorities determine, quite naturally, which questions you pursue and which questions you choose to leave aside.

This then is one major difference between the two proofs, having to do with their overall aims and interests. Another major difference has to do with the technical tools used to achieve those aims, especially ratios and proportions.

Once again, this difference may be expressed in simple quantitative terms: Archimedes' solution has many more proportion statements than Khayyam's. Of Archimedes' 40 steps, 16 assert proportions (40 percent); of Khayyam's 35 steps, only 4 assert

[60] Schramm (1965) n. 82.

proportions (11 percent). Instead of *proportions*, Khayyam more often asserts *equalities*, and he asserts 8 equalities in his argument. Of course, equalities are much less central to Khayyam's overall argument than proportions are to Archimedes', but this is because many of Khayyam's claims have to do directly with possibility or impossibility under various inequalities. Both proportions and equalities are backgrounded in Khayyam's treatment – relative to the study of cases – while they are both foregrounded in Archimedes' treatment. What we now see is that, *among* the two, Archimedes foregrounds proportions, while Khayyam foregrounds equalities. Archimedes' 16 proportions compare with 9 or 11 equalities: Khayyam's 8 equalities compare with 4 proportions.

Once again, the distinction between "foreground" and "background" is more qualitative than quantitative. As noted above, Archimedes has many geometrical constructions whose main function is to yield proportions – in particular, the grid of parallel lines, with its ensuing similar triangles. Khayyam has no need for such auxiliary structures and derives his relations in a much more direct way, from the equalities inherent in the conic sections; hence his much simpler figures.

Another example helps to bring forwards the sense of "foreground." I mentioned above the "9 or 11 equalities" Archimedes has: this is because some of his equalities are, as it were, self-effacing. Consider: "(9) it is: as the <line> EA to the <line> AΓ, so the <area> Δ to some <area> smaller than the <square> on BE, (10) that is, <smaller> than the <square> on HK." Now, the mathematical content of Step 10 is

$$(\text{sq. BE}) = (\text{sq. HK}),$$

but this is expressed through the "that is" operator, an after-thought to Step 9, so that, syntactically, we are invited to read Step 10 as a truncated way of stating

$$(\text{EA}:\text{A}\Gamma)::(\langle\text{area}\rangle\ \Delta:\langle\text{area}\rangle\ \text{smaller than sq. HK}).$$

Thus the equality is truly a background to the main statements, which are all about proportion. Put simply: for Archimedes, equalities are ways of getting at proportions while, for Khayyam, proportions are ways of getting at equalities. As in the issue of cases versus geometrical properties, the main question is which serves which. We may compare, for instance, the ways through which the two geometrical proofs reach their goals:

Archimedes: "(36) while the <rectangle contained> by ΣZN is equal to the <square> on ΣΞ, (37) that is to the <square> on BO, (38) through the parabola. (39) Therefore as the <line> OA to the <line> AΓ, so the area Δ to the <square> on BO."

Khayyam: "(28) So the ratio of the square of GC, the first, to the square of BC, the second, as the ratio of BC, the second, to GA, the fourth. (29) So the cube of BC – which is equal to the given number – is equal to the solid whose base is the square of GC, and its height GA."

Archimedes develops some equalities – only to translate them into proportions; Khayyam develops some proportions – only to translate them into equalities. The reason for this is, in fact, obvious: the way in which the goal is obtained is determined by the goal itself. Since the problem is set out by Archimedes as that of finding a proportion, it is to a proportion that his argument would lead; while Khayyam starts from an equality and must return to it. As it were, in the different melodies of their mathematical arguments, Archimedes has "proportion" as the tonic – the note from which he started and to which his readers expect him to return; while Khayyam has "equality" as the tonic.

In short, then, "proportion" gets foregrounded by Archimedes, "equality" by Khayyam. It is for this reason that Archimedes' lines are so clearly felt as "lines": a ratio involving four lines and areas, and ultimately dependent upon some geometrical similarity, is not easy to read off as a quantitative statement, but makes more sense as a qualitative statement about a geometrical object. This is true even of algebraically seeming statements, e.g., what we might express by a:b = ak:bk. Consider: "(18) and as the <line> ΓZ to the <line> ZN, (taking ZH as a common height) so is the <rectangle contained> by ΓZH to the <rectangle contained> by NZH." Inside a complex grid of lines, and inside a complex four-term expression, this claim becomes easier to read as a statement about figures in space, and not just about manipulated quantities. Khayyam's simpler equalities, on the other hand, are very easy to interpret as simple results of calculation, so that, even though his conception must have been thoroughly spatial, it becomes much more natural to read those equalities in abstraction from space – as it were, the equalities tend to become "equations."

We may sum up the comparison like this, then. Archimedes foregrounds geometrical properties, backgrounds study of cases; Khayyam foregrounds study of cases, backgrounds geometrical properties. Within geometrical properties, Archimedes foregrounds proportions, backgrounds equalities; Khayyam foregrounds equalities, backgrounds proportions. Put schematically:

Archimedes: (Proportions>Equalities)>Cases
Khayyam: Cases>(Equalities>Proportions).

It is this inverse ordering of foreground and background that makes the proofs so different, which finally makes us feel that Khayyam's proof "just couldn't be Greek" – that it is, indeed, already *algebra*. The mathematical materials are all the same, but they are arranged in a completely new structure. It is at this structural level, then, that Khayyam's originality has to be understood. In section 3.7 below, we shall try to explain the structural features of Khayyam's solution in terms of his mathematical practice. But first, another comparison is called for.

3.6 Khayyam's polemic: the world of Khayyam and the world of Archimedes

In a sense, Khayyam may be compared with the Greek author Archimedes: both offer a solution to the problem of proportion with areas and lines. In another sense, Khayyam may be compared with Greek authors such as Dionysodorus and Diocles: both Khayyam, and the Greek authors, did not merely offer a solution, but also (whether implicitly or explicitly) criticized previous solutions. Khayyam's style of offering polemics – no less than Khayyam's style of offering solutions – may provide us with a sense of Khayyam's mathematical practice. Let us then consider Khayyam's polemical treatment of the past.

For there is no question that Khayyam's treatment of the past was *polemical*. When we point out that Greek science was characterized by authors' desire to compete with their peers, as well as with their tradition, this does not mean that other authors were not competitive: most are. The truth is, the writing of books is often a difficult, dreary endeavor. To discover a solution might be pleasant; to proofread it prior to publication, is not. To produce books, then, one must have some special motivation for doing so which, emotionally, usually has to do with the desire to excel. The creative and the competitive are never distant. So much for the universals of human history. Passing to history itself, such general observations can gain meaning if we try to distinguish between the different modes that competition takes, historically. No culture is truly "eirenic," one in which scholars do not seek to compete with each other.[61] But different cultures lay down different rules for the tournament. This, then, has consequences for the intellectual practice that, ultimately, affects the intellectual contents themselves.

Let us then try to distinguish Khayyam's polemical style from that of Greek authors. And the first, surprising observation, is that, in a sense, Khayyam is more polemical: that is, polemics tends to be more explicit in his writings. We recall the interlude inside his solution (between Steps 20 and 21):

> This tangency or intersection was not grasped by Abu'l-Jud, the eminent geometer, so that he reached the conclusion that if BC is bigger than AB, the problem would be impossible; and he was wrong in this claim.
>
> And this kind is the one that baffled Al-Mahani (among the six kinds). So that you shall know.

Khayyam makes two explicit polemic statements in a sequence: (1) Abu'l-Jud made a specific geometrical mistake within his purported solution to the problem; (2) Al-Mahani failed to solve it. The second polemical claim, indeed, looms large in Khayyam's treatise. In Khayyam's introduction, the entire sequence of problems and solutions is seen to stem from Al-Mahani's statement, and failure to solve, this particular cubic problem. As for Abu'l-Jud, Khayyam, once again, does not leave this critical remark merely as

[61] It was customary to point to China as a possible "eirenic" culture: see Lloyd's qualification of that description in Lloyd (1996).

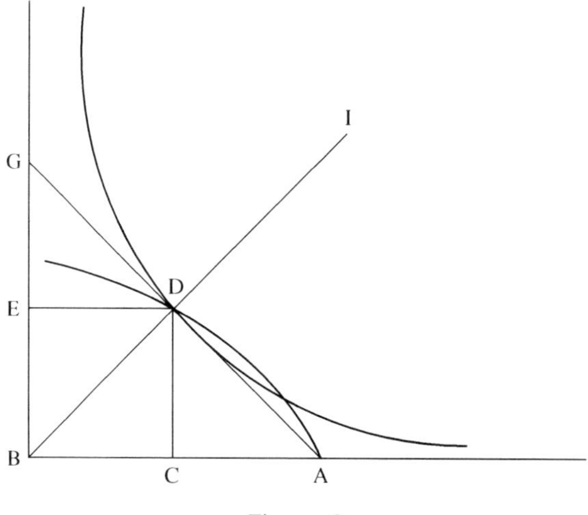

Figure 18

an aside between Steps 20 and 21. Having essentially completed the project of the *Algebra*, Khayyam went on to an appendix, criticizing Abu'l-Jud in detail for this geometrical mistake.

The structure of that appendix is as follows. Khayyam mentions that, five years after completing his *Algebra*, someone brought to his attention that Abu'l-Jud had already pursued a similar project of enumerating kinds of problems and solving them with conic sections. Crucially – Khayyam hastened to add – Abu'l-Jud's treatise was inexhaustive in two ways: it did not have all kinds, and it did not always treat the conditions of solubility. Still, a vexing discovery no doubt: and Khayyam set out to add the appendix (and, presumably, to add the references to Abu'l-Jud elsewhere in the treatise). After commenting on the incomplete nature of Abu'l-Jud's treatment, Khayyam then pounces upon a mistake made by Abu'l-Jud. The main issue is as follows:[62]

Says Abu'l-Jud: "We suppose that the number of the squares is the line AB. And we cut off from it the side of the cube equal to the number, that is BC. Then the line BC is either equal to CA, or greater than it, or less." He added: "If it is equal to BC, we complete the surface CE, we construct at D a hyperbola which

[62] I use here, with minor modifications, Rashed's English translation (Rashed [2000] 161), instead of paraphrasing the text again, as there is no need here to compare Khayyam's text with Archimedes'.

does not meet AB, BE, and we construct a parabola whose vertex is the point A, whose axis is AB and whose orthia is BC. Then the section will inevitably pass through the point D, as we have explained." He then pretends that the two sections touch each other at the point D. But he is mistaken, for they will necessarily intersect.

It is clear that Khayyam's text alternates direct quotation with comment and paraphrase. It appears therefore likely that Khayyam reproduces essentially Abu'l-Jud's own diagram, which is interesting: the diagram is almost identical with Khayyam's for this problem, with a minor difference in the labeling. (One wonders if Khayyam' solution, then, did not ultimately depend upon Abu'l-Jud.) It is also clear that Abu'l-Jud, in fact, never stated explicitly that the sections are tangent at D (Khayyam would certainly have quoted such a statement). Rather, it appears that this was an implicit assumption of Abu'l Jud. It is conceivable that a mathematician able enough to find a solution to the problem, could make such a mistake: as we have seen several times already in this book, the "topological" properties of conic sections are not very well covered in the Greek tradition. One can well imagine Abu'l-Jud deceived by his diagram. Conic sections would be drawn by the compass, as circular arcs. It is very tempting to draw the parabola as a quadrant with its center on B, and then the hyperbola is naturally drawn symmetrically, as a bigger arc, in the opposite sense to the "parabola," its center on the line BI (fig. 19). Then the tangency of the "conic sections" at D becomes a very persuasive illusion. (For an example of such representations of conic sections, consider fig. 20, based on the manuscript tradition for Menaechmus' alternative solution to the problem of finding two mean proportionals, as preserved by Eutocius.[63] Notice how the two lines passing through BZ, supposedly both conic sections, are represented by symmetrically intersecting arcs of circles).

So the ultimate reason why Abu'l-Jud did not notice his mistake is that, in all likelihood, *he had never stated it*. He did not say the sections were tangent at point D (though Khayyam could suspect, from Abu'l-Jud's diagram, that he had thought so). Was Khayyam's

[63] See Heiberg (1915) 82–4, where however the diagram is misleading.

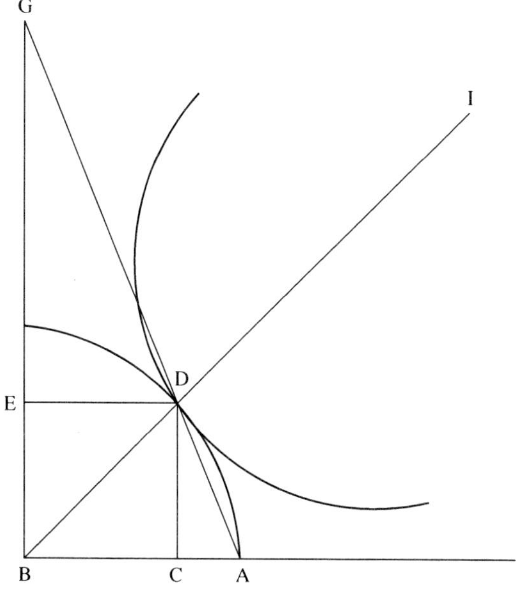

Figure 19

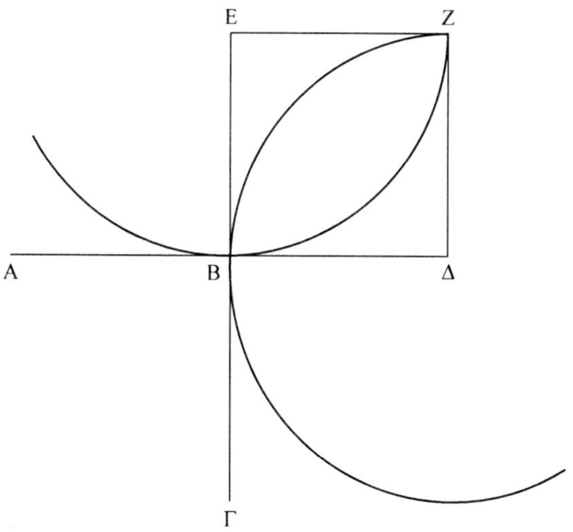

Figure 20

criticism then unfounded? Not so: Abu'l-Jud did commit the error, implicitly. We may quote from later on in Khayyam's appendix:

> And as to his statement: *if BC is greater than AC, the problem will be impossible because the two sections will not meet*, it is an unsound remark.[64]

This last comment by Abu'l-Jud was, in fact, the telltale sign: the mark left by Abu'l-Jud's evidence. Assuming (without stating so) that the sections touch at D, Abu'l-Jud went on to assert, dogmatically, that with BC>AC, they will not meet (moving even minimally "to the right of" tangency, one would obviously reach non-tangency). In other words: Abu'l-Jud's explicit mistake came not in the course of the geometrical development of the problem, but in the course of his treatment of the conditions of solubility. This, indeed, is precisely the way Khayyam portrays the situation. He does not criticize Abu'l-Jud for making a mistake as regards conic sections, but for making a mistake as regards the solubility of a cubic problem.

Following these quotations from Abu'l-Jud, Khayyam returns to the configuration of conic sections and discusses it again, showing the intersection (rather than tangency) at the point D. Typically, the discussion has an articulate structure: besides offering a general geometrical argument, Khayyam also offers numerical values with which the intersections can be followed. In this case, the numerical example serves an obvious persuasive aim, as part of the polemical argument: Khayyam is under a stricter requirement to *convince* his readers.

Turning now to compare Khayyam's polemics with those of Dionysodorus and Diocles, one immediately sees why Khayyam, unlike the Greek authors, could be so much more explicit in his criticism. He could be detailed in his criticism, because his treatment was directly comparable to that of his predecessor. There is very little continuity between the three treatments by Archimedes, Dionysodorus, and Diocles: they had made sure there would not be, by positioning their solutions as unique, independent results. But Khayyam, to start with, is engaged in arranging results according to strict principles of classification, so that it becomes straightforward

64 Rashed (2000) 162.

to position his results relative to previous ones. His, Khayyam's, treatise, covers all permutations of the four terms (cube, square, root, number); Abu'l-Jud covered only some. His, Khayyam's, treatise, studies all conditions of solubility; Abu'l-Jud studied only some. And finally his, Khayyam's, solution to *this* problem, recognized *this* (correct) condition of solubility, which Abu'l-Jud had missed. The most important feature is that it is possible to speak of *this* problem. With Archimedes, Dionysodorus, and Diocles, it takes considerable mathematical insight to perceive the basic identity of the three problems – which, in fact, is somewhat misleading: Dionysodorus' solution is somewhat less general than Archimedes', while Diocles' solution generalizes the problem relative to different parameters. But with Khayyam and Abu'l-Jud, the two authors solve – one correctly, the other not quite – the very same problem.

Why is that? There are several reasons. One, mentioned already, is that Khayyam's systematization of the structure of equations helps to identify the problem as equivalent to a particular, well-defined slot within the structure. Another is that the problem (much earlier in the history of Arab science – starting with Al-Mahani) had obtained a canonical formulation, in the terms of Al-Khwarizmi's algebra. Geometrical configurations, quite simply, can vary much more than can the terms of Al-Khwarizmi. Finally, the geometrical configuration itself is nearly identical in Khayyam's and Abu'l-Jud's treatments, so that Khayyam can easily pin down Abu'l-Jud's mistake.

This last fact is remarkable. How come Khayyam did not produce his own spectacular geometrical configuration – one comparable to the brilliance of, say, Diocles? Why did he keep to the minimal configuration of the parabola and the hyperbola, very probably already available to him from the tradition? But it is clear that, unlike Diocles, Khayyam did not aim to excel in his individual configurations. The solutions offered by Khayyam, taken separately, are not necessarily distinguished, and display little variability of conception. The aim is not to produce individually brilliant solutions, but to produce a system. The dimension of originality preferred by Khayyam is different from that of Diocles. To Diocles, it is important to obtain a solution original on its own; to Khayyam, it is important to obtain a solution original in its role within a system. It thus follows that Khayyam, unlike Diocles, would, on the one hand, aim to make his solutions simple and mutually comparable, and, on the other hand, would not aim to make them different from past solutions. So, finally, Khayyam – but not Diocles – could have his results directly comparable with those he criticized.

The Greek author brushes aside the results of past mathematicians, pretending to ignore them and suggesting that work should proceed again from scratch. The Arab author – whose modus operandi relies essentially on the collection and completion of past works – aims, instead, to subsume the results of past mathematicians within his own work. An interesting confirmation of this tendency can be seen not only in Khayyam's critique of past mathematicians, but in the critique of Khayyam himself by, possibly, his most impressive intellectual descendant, Sharraf Al-Din Al-Tusi.

The main work of Sharraf Al-Din Al-Tusi (late twelfth century) has suffered strange neglect.[65] Soon after its writing, it was revised and somewhat abbreviated by an anonymous author. No manuscripts of the original, unabridged text, survive. A late thirteenth-century manuscript of the abridgment lay unnoticed in Patna, India; an eighteenth-century manuscript formed part of the British India Office collection in London, where it attracted no more than passing attention from scholars. The first modern edition of the work, produced thanks to the labors of Rushdi Rashed (who also discovered the Patna manuscript) came out only in 1986. This contains possibly the most remarkable achievement of early Mediterranean mathematics. I shall not go here into the detail of Al-Tusi's achievement, and shall merely point out the direction his work was taking: this is telling for the nature of development in Arabic science.[66]

We are not sure of the title of Al-Tusi's main work: the anonymous reviser had supplied the title "The Equations," though an original title "Algebra" is as likely. The work is very obviously in

[65] Thus, one must be warned not to confuse Sharraf Al-Din Al-Tusi with his much better known namesake, Nasir Al-Din Al-Tusi.

[66] The reader should consult Rashed (1986) for fuller details on Al-Tusi and his text. Hogendijk (1989) is useful for offering a mathematical intepretation of Al-Tusi that keeps to within the conceptual tools available to Arabic mathematicians.

the tradition of Khayyam's *Algebra*. The approach taken is much more impersonal than Khayyam's: there are very few explicit bibliographic statements made through the work, and certainly no explicit polemic. (Possibly, however, this may be due to the abridgment of the work.) The main intention is clear, however: to reformulate the entire system of Khayyam's *Algebra*. The primary transformation consists in a new ordering principle. Khayyam proceeded through a geometrical principle, reflecting the main division he saw in the tradition available to him. He divided the various equations into kinds, primarily, according to the geometrical tools they required. Especially, some had required conic sections, and some did not: this was the main division of Khayyam's work. Throughout, however, Khayyam had applied a different classification: do the problems admit of solutions without limits, or not? This was not the major dimension of classification, and so was somewhat blurred in Khayyam's system. There was no obvious principle by which one could expect this equation, but not the other, always to have a solution. On the other hand, this distinction was often very important to Khayyam. Paradoxically, just because this division was not built into the system of classification, it became marked by Khayyam's repeated assertion that this problem or the other did not always admit of solution.

The natural way forwards for Al-Tusi, then, would be to improve on Khayyam's classification by basing it on this, apparently deeper principle. His main division is into equations that are always soluble, and those that are not. He has twenty equations that do not have limits of solubility, and five that do. His treatise culminates with a very thorough study of those five equations: the first of which, indeed, is Archimedes' problem itself.

The change in order could be merely cosmetic, but the point is deeper: once you group together the equations that have limits on their solubility, it becomes natural to systematize the search for such limits. In that Al-Tusi differs from Khayyam. Khayyam comes across limits on possibility haphazardly: his one advantage over Abu'l-Jud is that he does make sure he would come across *all* of them. Since Khayyam does not look for a principle uniting all such limits on possibility, he has each case of the limits of possibility dependent upon the local terms adopted in the local geometrical solution. In the case of Archimedes' problem, then, Khayyam's statement of the limits of possibility derive directly from the configuration of hyperbola and parabola (so that, for instance, he does not notice explicitly the value of the maximum as one-third the line). Al-Tusi, however, by the very nature of his system, is led to look for the conditions of solubility as such: he foregrounds those conditions and allows them to dominate his search for solutions. He thus starts his solution of Archimedes' problem from the observation that the limit occurs at one-third the line, which he then derives from the maximum of the associated solid at that point. Since this is the first result obtained – and since Al-Tusi by now comes from a mathematical tradition that is perfectly fluent in algebraical manipulations – Al-Tusi studies this property without having recourse to conic sections. This time, he obtains this result by direct calculation on equations. This then becomes the pattern for the remaining equations that have limits on their solubility. For each, Al-Tusi first constructs a geometrical object associated with the equation, for which a maximum is obtained at a certain point; he proves the existence of the maximum, and in this way shows the existence of the limits on solubility. But wait: the maximum is also a limiting case, where the equation admits of only one solution. Al-Tusi proves that this is the case, and then – most remarkably – he proceeds to deduce the values of the *pairs* of solutions, away from the maximum, as a function of the value obtained at the maximum. (For an example of this difficult algebraic exercise, see e.g., Rashed [1986] II.5–8 for the finding of the pair of solutions for Archimedes' problem.)

Perhaps the word "polemical" is inappropriate: we do not know Al-Tusi's feelings about Khayyam. What seems clear is that Al-Tusi set out to produce a work that excels in comparison with Khayyam. And the point is that, to do so, he produced a work that is very directly comparable to Khayyam. Al-Tusi covers, visibly, the same ground as Khayyam. How can he excel, then? Merely by improving the structural properties of the work and by extending the results at least for some elements within the structure. The nature of polemic determines the nature of desired evolutionary process. The nature of polemic in Classical Greek mathematics favored the evolutionary process where more and more elegant solutions, to essentially the same problem, are accumulated. The nature of polemic in late ancient and medieval mathematics favored the improvement where more and more complete and well-structured theories are formed. The principle guiding the development of Arabic mathematics is thus the typically modern one, to extend and to subsume. From Abu'l-Jud to Khayyam, and then from Khayyam to Sharraf Al-Din Al-Tusi, we see two iterations of this operation of extension and subsumption, the result being an enormous growth in the power of mathematical techniques. That no further iterations were made by Arab mathematicians is to a large degree a function of historical forces extrinsic to mathematics. Or, better put: when European mathematicians begin their exponential process of extension and subsumption, in the sixteenth and seventeenth centuries, they already start from a tradition much richer than that of Classical Greek mathematics. In this respect, modern European mathematics is of a piece with its Arabic ancestor.

Such meta-historical considerations aside, it is clear that, with Al-Tusi, we seem to breathe a distinctly modern air. Here is a systematic study of maxima and the functional relations between solutions. But note: we have already seen a glimmer of such a systematic study in Eutocius' brief comments on Archimedes' and Diocles' solutions. I have suggested, in section 2.4 above, that the awareness of functional relations between mathematical objects could be the product of the deuteronomic practices of late ancient and medieval mathematics. One needs the genius of Al-Tusi (and the example of Khayyam) to reach as far as Al-Tusi did: but the basic transformation of mathematics is due to the mathematical practice itself.

To see this, let us now wrap up out discussion of Arabic mathematics, by returning to Khayyam's work itself: what, ultimately, made it so different from Archimedes?

3.7 How did the problem become an equation?

In sections 3.5–6 we have seen two ways in which Khayyam values systematicity. In section 3.5, we saw how Khayyam, inside his own work, was interested in those features that highlight the properties of individual solutions as belonging to a system. In section 3.6, we saw how Khayyam (as well as Al-Tusi following him) was interested in those features that made his work systematize previous solutions. Section 3.5 focused on internal systematization – the interrelation of Khayyam's solutions to each other – while section 3.6 focused on external systematization – the interrelation of Khayyam's solutions to previous ones. The two are of a piece: because Khayyam worked in an environment where scientific excellence was understood in the terms of better arranging results available from the past, he was driven to favor the systematic features of solutions over the elegance of each individual solution, with the resulting characteristics of his algebra.

Let us consider again these characteristics. What, finally, makes Khayyam's propositions into solutions of equations, rather than of geometrical problems?

In several ways, Khayyam downplays geometry. Most obviously, the foregrounding of the study of cases is a feature of the work at all levels – the overall treatise as well as the individual proof. Throughout, Khayyam is motivated by the impulse to provide exhaustive lists. And it is because this proof serves as a "case," that it is analyzed according to its cases. Khayyam, as it were, never really set out to solve a problem – this was not the issue. The issue, for him, was to catalogue a certain problem according to the properties of its solution.

So much for the foregrounding of the study of cases over geometrical properties. Inside geometrical properties, once again though in a less obvious way, the foregrounding of equalities over proportions is determined by the overall impulse to provide exhaustive lists. Equalities lend themselves to an exhaustive survey; proportions do not. Equalities have the simplest possible surface structure: a pair of symmetrical positions. Proportions have four positions, symmetrical in some ways and asymmetrical in others. Also, subtraction can always be eliminated from equalities (instead of $A = B - C$, you can have $A + C = B$), but not from proportions, lending a further dimension of complexity to proportions. Those brute facts alone make it almost inevitable that, when motivated by a desire to provide exhaustive lists of mathematical relations, equalities will be foregrounded over proportions.

On the other hand, Archimedes is throughout motivated by immediate geometrical tasks – in this case, to divide a sphere according to a given ratio. This ultimate goal determines the nature of Archimedes' treatment, just as Khayyam's exhaustive list determines his own treatment. Archimedes foregrounds the solution with its specific geometrical property, because this geometrical property is the external function of the proof; and he foregrounds proportions, because this external function is ultimately determined by a ratio. True, proportions are not easy to catalogue, but Archimedes was never interested in cataloguing his problem. In his treatment, Archimedes' problem seems to be a one-off, totally unrelated to any other problem. Archimedes is simply interested in obtaining interesting geometrical tasks, and obtaining results according to given ratios is often an interesting task, *just because* proportions are more complicated. In his same treatise, the *Second Book on the Sphere and Cylinder*, Archimedes does mention, of course, tasks involving simple equalities. There is the task to find a plane equal to the surface of a given sphere; or to find a sphere equal to a given cone (or cylinder). But those problems are, for Archimedes, absolutely trivial. The first does not even get a diagram, and is effectively dismissed as obvious from the facts known from known results; the second gets a brief treatment in the first proposition of the book, where the proof, once again, is a mere quick unpacking of well-known results.[67] The remainder of the treatise is then dedicated to real problems, which are all defined by proportion or by the (equivalent) relation of similarity.

Archimedes' problem arises, as it were, in "real-life geometry," and its shape is determined by the demands of this "real-life geometry." Khayyam's problem arises from its position in a list of problems – the list deriving not from an external, geometrical investigation, but from its own independent listing principle.

This comparison is crucial. There are of course Classical Greek mathematical texts that list results: these are known as "Elementary results." But the essence of Greek elementary results – the very way in which Greeks understood what "elementary" means[68] – is that

[67] For these two problems, see Heiberg (1910) 170–4.
[68] The *locus classicus* is Aristotle, *Metaph.* 1014a35–b2.

those results serve in some other, advanced situations. Thus, Book II of the *Elements* – once considered as an example of "geometrical algebra" – is in fact motivated by an interest in specific geometrical configurations, arising in specific advanced problems. Saito has shown that the results of Book II are arranged not according to some principle internal to the work itself, but by geometrical motivations that are external to it.[69] Thus it is only natural that no attempt is made, in Book II, to obtain anything like an exhaustive list. The list is not interesting for its properties as a list, but is a mere *repository* of results, useful case by case.

The very same problem, we see, may be set in very different types of context. The Archimedean context is, as it were, "vertical": the problem of finding lines satisfying a certain ratio is not related to other problems of lines satisfying certain ratios, but is related to a different *kind* of problem, that of cutting a sphere. The Khayyamite context is horizontal: the problem of finding lines satisfying a certain equality is not related to other problems from which it may arise, but is instead related to other problems of finding lines satisfying other equalities. This difference in context fully determines the mathematical difference between Archimedes and Khayyam. Khayyam differs from Archimedes in his foregrounding of study of cases, and of equalities, both deriving from his different type of context. Thus, merely by being set in different types of context – with no deep difference in admissible mathematical operations – the very nature of the proposition has been transformed, and a geometrical problem has become a cubic equation.

The question arises, why does Khayyam's context differ so markedly from that of Archimedes. And, in a sense, we already have been given a possible answer to this question. When surveying the overall structure of Khayyam's treatise, we saw that the impulse to provide exhaustive lists is closely related to a basic feature of the work, namely the continuity it displays between introduction and discussion. General, meta-mathematical claims, are interspersed with more specific mathematical claims at the object level, and the claims at the object level gain their significance from the claims at the meta-mathematical level. It is because Khayyam

[69] Saito (1985).

is primarily interested in positioning his work in the context of the past, that he is focused on the idea of an exhaustive list. The treatise thus has an essentially second-order character. It is an unending introduction; it does contain, to be sure, many problems set out and solved – but it is considered throughout not *through* problems, but *about* problems.

In other words, the main difference between Archimedes and Khayyam is that, whereas Archimedes separates clearly his introductions from his main text – and uses them, so to speak, merely as introductions – Khayyam does not separate his general claims at all from his actual mathematics, and allows the general claims, instead, to govern the particular claims.

Now, to produce, for the first time, an exhaustive list of equations up to the third degree, and to solve them all, and to achieve all this with great elegance and precision, is a task calling for genius. Thus it is not as if the basic stylistic difference between Archimedes and Khayyam *explains* Khayyam's treatise. No one letting his introductions run wild would write Khayyam's *Algebra*.

But while this stylistic difference does not provide sufficient conditions for the writing of the *Algebra*, it does provide, I argue, a central necessary condition. For the *Algebra* to be written, one needed first of all to have a culture where writing *about* mathematics was part and parcel of the writing *of* mathematics.

The urge to arrange together the achievements of the past runs through late ancient and medieval mathematics. In Arabic mathematics, and concentrating on the history of Archimedes' problem of a proportion with areas and lines, we can see how this urge gradually transforms the problem. We may start with Al-Khwarizmi, who creates the syncretistic language of algebra where calculation and geometry are made equivalent. (This is the urge to create relations between the components of the tradition.) We may then follow Al-Mahani's canonical statement of the problem, respected by later Arabic authors, so that the problem gains a well-defined meaning as a single equality with cubes. (This is the urge to keep to the standard forms of the tradition, so as to have your own results directly comparable to those of the past.) Finally, the same urges – to systematize and to make your work comparable with the past – leads authors such as Khayyam and Al-Tusi to inscribe the problem within rational *lists* of problems and solutions. This foregrounds non-geometrical features of the problem, definitely turning it into an equation.

We have come back to Klein's original insight: equations differ from problems, in that they are somehow second order. They are not directly defined by the objects to which they refer, but by their systematic interrelationship with other, similar mathematical expressions. We have offered, however, a different historical account of the route leading from Greek problems to Arab equations, passing through the deuteronomic culture of late antiquity and the Middle Ages. As a conclusion, let us review this historical account.

CONCLUSION

The outline of the argument of the book has now been repeated several times. Hellenistic Greek mathematical practice focused on the features of the individual proof, trying to isolate it and endow it with a special aura. Thus the characteristic object of Hellenistic Greek mathematics is the particular geometrical configuration. Medieval mathematical practice focused on the features of systems of results, trying to bring them into some kind of order and completion. Thus the characteristic object of medieval mathematics is the second-order expression. In a particular geometrical configuration, the mathematician foregrounds the local, qualitative features of spatial figures. In a second-order expression, the mathematician foregrounds the global, quantitative features of mathematical relations. Thus, Hellenistic Greek mathematics – the mathematics of the *aura* – gave rise to the *problem*; medieval mathematics – the mathematics of *deuteronomy* – gave rise to the *equation*.

The comparison between the two kinds of mathematics is at its starkest when we compare Hellenistic Greek mathematics directly with advanced Arabic mathematics. This comparison is useful, then, to get a sense of the nature of the transformation. But, to look for the historical account for this transformation, we have concentrated in this book on a more subtle comparison. In this book, I have given much attention to the transitional stage of Late Antiquity, already different from Classical Hellenistic mathematics, though in ways that are less obvious. In the work of Eutocius, we saw suggestions of the direction ahead. We came across expressions treating geometrical relations as if they were multiplications and additions; and expressions describing the functional relations between points on a curve. Such expressions are rare in Eutocius: they are also unimaginable in Hellenistic Greek mathematics – and natural in advanced Arabic mathematics.

No one can ascribe to Eutocius any deep originality as a thinker. To find in him, already, the characteristic features of medieval mathematics, is therefore remarkable. But once we see that those features arise not from conceptual developments, but from changes in mathematical practice, Eutocius' originality becomes clear. Eutocius' mathematics was already different, in terms of its practice, from that of Hellenistic Greek mathematics. Archimedes looked for striking results standing on their own; Eutocius looked for systematization. Hence the new features of Eutocius' mathematics. Later on, to get from Eutocius to truly remarkable systematizations, what was required was mostly an added mathematical ambition – of which Arabic mathematicians had plenty.

The book concentrated on a single thread of history – the transformation of Archimedes' problem of the proportion with areas and lines. The choice was, of course, to some extent arbitrary: it was simply useful that the problem has attracted enough attention so that a history could be written surrounding it. But this problem also has an added important property: it is complicated. Complexity, in itself, is an engine of change, for the following reasons. First, as we recall, already in the writings of Archimedes himself one could detect the signs of tension. The problem seemed to defy conventional approaches. Thus, the statement of the problem was general and seemingly quantitative; Archimedes' language suggested the treatment of geometrical relations as multiplication and addition. All of this could be accounted for, and coincided with Archimedes' fundamental geometrical conception of the problem, typical of his era. Yet, historical change fastens upon such tensions: so that, for instance, Eutocius had to deal somehow with the gap between Archimedes' geometrical intentions, and his quasi-calculatory language, coming up with something approaching a reduction of geometry to calculation. We can put this in a nutshell: one reason why things in history do not stay the same is that they never were thus. Objects such as Archimedes' problem are rich in dialectical possibilities, and, when restated by later authors, they would have to be transformed. This is important, because we may often be tempted to view cultures as monoliths, ignoring the tensions inherent within past artifacts, generalizing about "the Greeks" and "the moderns" – as, indeed, I often have done myself in this book. But even while making generalizations about past cultures, historians should be aware of the ways in which past artifacts, inevitably, run against the limits of their culture. This is important not because "one should not over-generalize" (why not?), but because the dialectical, complex nature of the past is one of the reasons why it changed at all – why the past has given rise to *history*.

Such historiographical considerations aside, then, I suggest that the tensions inside a complex object are one reason why, in the history of mathematics, complexity is an engine of change. Another reason is that complex objects are many sided. The object has many features, which different authors may choose to foreground or background. Archimedes' problem can be solved, geometrically, only with conic sections; it involves, in its statement, a proportion of areas and lines. It gives rise to a difficult solution; it also has the property that it has a limit on its solubility. Because it is many sided, the problem could change, locally as well as globally. Let us start with a late example of a local change: Khayyam's treatment foregrounded the questions of solubility with or without conic sections, positioning the equation – still! – within a system of problems; Al-Tusi, later, foregrounded the property of the limit on solubility, positioning the problem within a group of studies of maxima. Khayyam did not even calculate the maximum for the problem; Al-Tusi used that maximum to derive the solution itself. Again, considering the global level, we can say that while Hellenistic Greek mathematicians foregrounded the role of conic sections, so that they saw the problem as an opportunity to construct configurations of such curves, medieval mathematicians foregrounded the statement of the problem as a proportion of areas or lines or, in other words, as an equality involving cubes – with the resulting algebraic canonical form obtained by Al-Mahani.

Here, finally, we reach what may be the philosophical explanation for why mathematics should have a history. True, the truths about mathematical objects can be equivalently stated in different languages, and so the change of language itself need not, directly, change mathematics. Thus, in principle, cultures may change, and mathematics remain the same. But it is rare that different cultures have the same mathematics, simply because different cultures would normally study different mathematical *objects*. A mathematical object is characterized by its features. Different cultures, speaking seemingly about the same object, would usually foreground and background different features of it and so, for all intents, would speak about a different mathematical object.[1] What you choose to foreground or background is highly sensitive to your cultural practice, and so it is simply unlikely to be invariant to culture. In truth, mathematicians do not necessarily *know* what it is they foreground and background: they simply happen to follow, systematically, this type of question and not that, to be bothered by this type of concern and not that. All those are issues of practice, not of any philosophical conception. A change of practice, then, will inevitably tend to change the mathematical object itself, regardless of what may happen, simultaneously, in terms of philosophical conception.[2]

We have now foregrounded the concept of foregrounding, and so I find it useful to restate my basic historical formula. Hellenistic Greek mathematics, whose practice may be summed up by the *aura*, foregrounded the local characteristics of configurations, giving rise to the *problem*; medieval mathematics, whose practice may be summed up by *deuteronomy*, foregrounded the global characteristics of relations, giving rise to the *equation*.

Looking back on the arguments made by previous historians of mathematics, we can therefore side with Unguru and Klein: there was a basic divide separating ancient, from later mathematics, typically seen in the transformation from a more geometrical approach to a more algebraic approach. Klein is also correct in that the very mathematical object seems to change. From what may be called

[1] Notice that this observation is independent of questions in the philosophy of mathematics as regards the nature of the mathematical object. Even if mathematical objects exist apart from the mathematicians who study them – which I actually believe they do – two mathematicians could appear to speak about the same entity and yet foreground radically different features of it, so that their theories, in fact, are about different objects. One studies the tail of the dog, the other its ear, and so one is, in truth, a tail-theoretician and the other an ear-theoretician: and yet dogs, tails, and ears may all exist, apart from the mathematicians, in a Platonic heaven. (Of course, in a non-Platonist account of mathematical objects, it is trivial that the objects change with the change in mathematical foregrounding.)

[2] This understanding of the nature of historical change – as driven by changes in the mathematical object itself, as a result of the different questions with which the mathematicians address their subject-matter, ultimately explicable in terms of mathematical practice – is in line with several recent studies in the history of mathematics, e.g., Goldstein (1995), Corry (1996), Herreman (2000).

a first-order mathematical object, studied by ancient mathematicians, we pass to what may be called a second-order mathematical object, studied by the moderns. But Klein, I believe, got the history wrong: the transformation was earlier – and more gradual – than Klein believed; and it was driven not by a direct change in the philosophical conception of the mathematical object but, indirectly, by the change in mathematical practice, leading to a different foregrounding of the features studied within mathematical objects.

The notion of foregrounding allows us, finally, to discuss briefly not only the notion of mathematical change but also that of mathematical *progress*.

It is obvious that mathematics has made enormous progress from early Mediterranean mathematics to modern times. But this notion of progress is problematic for anyone familiar with early Mediterranean mathematics. The works of Archimedes, or of Khayyam – to take two clear examples – are simply of the very highest intellectual quality. In a sense, they are clearly inferior to today's first year university textbooks but, in another sense, they are also clearly superior to them. This, then, seems to be a puzzle, putting into question the notion of mathematical progress.

Two observations made in this book go some way to resolve this puzzle. First, I noted the specifically historical character of the drive of advanced Arabic mathematics – to return to earlier works and to extend and subsume them. This is a powerful mechanism of iteration that has the potential for enormous growth in mathematical techniques – leading, in the case mentioned in section 3.5 above, from Abu'l-Jud to Sharraf Al-Din Al-Tusi. The same iterative mechanism was active, for somewhat different reasons, in modern European mathematics. Qualitative growth through time is thus one of the inherent properties of this type of mathematics. External reasons, finally, such as the political encouragement for creating very large groups of mathematical practitioners, has made this growth much more powerful than in any period in the past. This combination makes modern European mathematics unique. When we castigate non-modern mathematics for failing to become as powerful as modern mathematics, we take a certain mathematical practice for granted, as if the mere application of intelligence should give rise, through time, to qualitative growth. But this qualitative growth through time is, I argue, just as historically contingent as any other property of mathematical practice. To have it, you do not have to be intelligent: you have to be modern.

Most important, however, is that the notion of foregrounding reminds us that different kinds of mathematics, quite simply, aim at different goals. In our histories, mathematicians are typically judged by the standard of modern mathematics, and the focus is on the contribution of the individual work of mathematics to the entire body of mathematical knowledge and techniques. Hellenistic Greek mathematicians then get, on the plus side, a mention for their achievements – deductive geometry, conic sections, a few other interesting beginnings – which is then offset by a mention, on the negative side, of all they failed to accomplish: algebra, analytic geometry, the calculus, etc. All of this is valid, and calls for explanation: I have tried, in the above, to suggest a possible outline for explaining the greater achievements of modern mathematics.

But then, Hellenistic Greek mathematicians did not aim to create algebra, analytic geometry, or the calculus. They set out to provide elegant solutions to problems, such that would be, individually, accomplished works of art. This they certainly achieved.

Viewed from the perspective of the goals of the present, it is only natural that past history appears like progress. But seen from the perspective of Greek mathematics, can we really say that the average contemporary proof compares, in elegance, to the Greek one? The point is deeper than just that of mathematical virtuosity. A modern mathematical proof, even when brilliant and polished, serves a wider purpose. A Greek solution to a problem may truly be a work in its own right. A Greek proof is, by its nature, a work of art; a modern one is, by its nature, a tool. And so, something of the aura of mathematics *has* been lost, in the transition from Greek to medieval – and modern – mathematics.

Mathematical progress is a reality: that is, the modern growth of the body of mathematical knowledge and techniques did happen. It is also legitimate and natural to value this growth. But I shall say just this: that there is value in the mathematics of past cultures, one which mathematical progress would not erase. This is one of the reasons to study the history of mathematics.

REFERENCES

Arafat, W. and Winter, H. J. J. 1950. The Algebra of Umar Khayyam, *Journal of the Royal Asiatic Society of Bengal, Science* 16: 27–78.

Benjamin, W. 1968. The Work of Art in the Age of Mechanical Reproduction, translated in *Illuminations* (ed. H. Arendt). New York. (Original: Das Kunstwerk im Zeitalter seiner technischen Rerodukzierbarkeit, *Schriften* 1: 366–405, ed. T. Adorno, 1955: Frankfurt.)

Clagett, M. 1964–84. *Archimedes in the Middle Ages* (5 vols.). Philadelphia.

Corry, L. 1996. *Modern Algebra and the Rise of Mathematical Structures*. Boston.

Cuomo, S. 2000. *Pappus of Alexandria and the Mathematics of Late Antiquity*. Cambridge.

Decorps-Foulquier, M. 2000. *Recherches sur les Coniques d'Apollonius de Pergé et leur commentateurs grecs. Histoire de la transmission des Livres I–IV*. Paris.

Fowler, D. H. F. 1999. *The Mathematics of Plato's Academy* (2nd ed.). Oxford.

Fried, M. and Unguru, S. 2001. *Apollonius of Perga's Conica: Text, Context, Subtext*. Boston.

Gandz, S. 1936. The Sources of Al-Khowarizmi's Algebra. *Osiris* 1: 263–77.

Goldstein, C. 1995. *Un Théorème de Fermat et ses lecteurs*. Saint-Denis.

Herreman, A. 2000. *La Topologie et ses Signes*. Paris.

Heath, T. L. 1897. *The Works of Archimedes*. Cambridge.

1921. *A History of Greek Mathematics*. Oxford.

Heiberg, J. L. 1885. *Euclides: Opera. Vol. IV*. Leipzig.

1891. *Apollonius: Opera. Vol. I*. Leipzig.

1893. *Apollonius: Opera. Vol. II*. Leipzig.

1898. *Ptolemy: Syntaxis. Vol. I*. Leipzig.

1910. *Archimedes, Opera Omnia, Vol. I*. Leipzig.

1913. *Archimedes, Opera Omnia, Vol. II*. Leipzig.

1915. *Archimedes, Opera Omnia, Vol. III*. Leipzig.

Hogendijk, J. P. 1989. Sharaf Al-Din Al-Tusi on the Number of Positive Roots of Cubic Equations, *Historia Mathematica* 16: 69–85.

Hoyrup, J. 1996. 'The Four Sides and the Area'. Oblique Light on the Prehistory of Algebra, in *Vita Mathematica* ed. R. Calinger, Washington DC, pp. 45–65.

Hughes, B. B. 1989. *Robert of Chester's Latin Translation of Al-Khwarizmi's Al-Jabr*. Stuttgart.

Jones, A. 1986. *Pappus of Alexandria: Book 7 of the Collection*. New York.
Karpinski, L. C. 1915. *Robert of Chester's Latin Translation of the Algebra of Al-Khowarizmi*. New York.
Klein, J. 1934–6/1968. *Greek Mathematical Thought and the Origins of Algebra*. Cambridge MA.
Knorr, W. 1982. Observations on the Early History of the Conics, *Centaurus* 26: 1–24.
1986. *The Ancient Tradition of Geometric Problems*. Boston.
Lloyd, G. E. R. 1996. *Adversaries and Authorities*. Cambridge.
Lorch, P. 1989. The Arabic Transmission of Archimedes' Sphere and Cylinder and Eutocius' Commentary, *Zeitschrift für Geschichte der Arabische-Islamischen Wissenschaften* 5: 94–114.
Mansfeld, J. 1994. *Prolegomena: Questions to be Settled before the Study of an Author, or a Text*. Leiden.
1998. *Prolegomena Mathematica from Apollonius of Perga to the Late Neoplatonists, with an Appendix on Pappus and the History of Platonism*. Leiden.
Mugler, C. 1972. *Archimède, Opera*. Vol. IV. Paris.
Musharrafa, A. and Ahmad, M. A. 1939. *Al-Khwarizmi: Kitab Al-Jabr wa Al-Muqabala*. Cairo.
Netz, R. 1998. Deuteronomic Texts: Late Antiquity and the History of Mathematics, *Revue d'Histoire des Mathématiques* 4: 261–88.
1999. *The Shaping of Deduction in Greek Mathematics: a Study in Cognitive History*. Cambridge.
2000. Why did Greek Mathematicians Publish their Analyses? In *Memorial Volume for Wilbur Knorr*, eds. H. Mandel, J. Moravcsik, and P. Suppes, Stanford, pp. 139–57.
2001. Archimede. *Encyclopedia Italiana della Storia della Scienza, Vol. I La Scienza Antica:* 779–790. Rome.
Forthcoming. *The Works of Archimedes Translated into English, Volume I: The Two Books on Sphere and Cylinder*. Cambridge.
Netz, R., Saito, K. and N. Tchernetska. 2001. New Reading of *Method Proposition 14: Preliminary Evidence from the Archimedes Palimpsest (Part 1), Sciamvs* 2: 9–29.
Rashed, R. 1986. *Sharaf Al-Din Al-Tusi: Oeuvres Mathématiques. Algebre et Géometrie au XIIe siècle*. Paris.
2000. *Les Catoptriciens Grecs: Les Mirroirs ardents*. Paris.
Rashed, R. and Vahabzadeh, B. 1999. *Al-Khayyam Mathématicien*. Paris.
2000. *Omar Khayyam the Mathematician* (partial translation of [1999]). New York.
Rome, A. 1936. *Commentaires de Pappus et de Théon d'Alexandrie sur l'Almageste, Vol. II*. Biblioteca Apostolica Vaticana: Vatican City.
Rosen, F. 1831/1969. *The Algebra of Muhammed ben Musa*. New York.
Saito, K. 1985. Book II of Euclid's Elements in the Light of the Theory of Conic Sections, *Historia Scientiarum*: 31–60.

Schramm, M. 1965. Steps Towards the Idea of Function: a Comparison between Eastern and Western Science of the Middle Ages, *History of Science* 4: 70–103.

Thomas, I. 1941. *Greek Mathematical Works*. Vol. II. Cambridge MA.

Saito, K. 1997. Index of the Propositions Used in Book VII of Pappus' Collection, *The Journal of Humanities, Faculty of Letters, Chiba University* 26: 155–8.

Schöne, H. 1903. *Hero: Opera. Vol. III*. Leipzig.

Sezgin, F. 1974. *Geschichte des arabischen Schriftums, Bd. 5: Mathematik Bis ca. 430 H.* Leiden.

Sesiano, J. 2002. Towards Omar Khayyam, *Farhang* 14: 193–201.

Toomer, G. J. 1976. *On Burning Mirrors/Diocles*. New York.

Unguru, S. 1975. On the Need to Rewrite the History of Greek Mathematics, *Archive for the History of Exact Sciences* 15: 67–114.

1979. History of Ancient Mathematics: Some Reflections on the State of the Art, *Isis* 555–65.

Vitrac, B. 1998. *Euclide/Les Éléments*. Vol. III. Paris.

Woepcke, F. 1851. *L'Algèbre d'Omar AlKhayyâmî*. Paris.

Youschkevitch, A. P. 1976. *Les Mathématiques arabes*. Paris.

Zeuthen, H. G. 1886. *Die Lehere von den Kegelschnitten im Altertum*. Kopenhagen.

INDEX

Lightning Source UK Ltd.
Milton Keynes UK
UKOW01n1855161116
287840UK00001B/1/P

9 780521 829960